Science
and
Technology Resources

Recent Titles in
Library and Information Science Text Series

Science and Technology Resources

A GUIDE FOR INFORMATION PROFESSIONALS AND RESEARCHERS

James E. Bobick
and
G. Lynn Berard

Library and Information Science Text Series

LIBRARIES UNLIMITED

AN IMPRINT OF ABC-CLIO, LLC
Santa Barbara, California • Denver, Colorado • Oxford, England

Library of Congress Cataloging-in-Publication Data

Bobick, James E.
    Science and technology resources : a guide for information professionals and researchers / James E. Bobick and G. Lynn Berard.
        p. cm. — (Library and information science text series)
    Includes bibliographical references and index.
    ISBN 978-1-59158-793-4 (hardback) — ISBN 978-1-59158-801-6 (paperback) — ISBN 978-1-59158-794-1 (ebook)   1. Science and technology libraries—Reference services. 2. Science—Bibliography—Methodology.   3. Scientific literature—Bibliography—Methodology.   4. Technology—Bibliography—Methodology.   5. Technical literature—Bibliography—Methodology.   I. Berard, G. Lynn.   II. Title.
    Z711.6.S35B63   2011
    025.5'2765—dc22   2011000461

ISBN: 978-1-59158-793-4
        978-1-59158-801-6 (pbk.)
EISBN: 978-1-59158-794-1

15   14   13   12   11   1   2   3   4   5

This book is also available on the World Wide Web as an eBook.
Visit www.abc-clio.com for details.

Libraries Unlimited
An Imprint of ABC-CLIO, LLC

ABC-CLIO, LLC
130 Cremona Drive, P.O. Box 1911
Santa Barbara, California 93116-1911

This book is printed on acid-free paper ∞

Manufactured in the United States of America

To
Sandi and Alan
We owe you a lot!

Contents

Preface

Science knows no country, because knowledge belongs to humanity, and is the torch, which illuminates the world.

—*Louis Pasteur*

Science happens everywhere. It's a part of our everyday lives. If we just become observant about our surroundings, we will begin to be aware of the magnitude of its importance. Science and technology advances, and the acquisition of knowledge necessary for those advances, rely on libraries and information centers for archiving the history of these fields and for providing materials essential for the provision of insights and ideas that result in new scientific developments and inventions and for a clear understanding of natural and material phenomena. The fields of science and technology are ones of rapid change and present challenges in staying abreast of new developments and discoveries. In the last millennium many new science and technology fields have been born. Science is fascinating and fast-paced. There is much to learn of traditional science knowledge while watching new developments spring up each and every day. Carefully built library collections staffed by highly skilled and well-trained professionals play a vital role in the advancement of knowledge about the physical and biological world we live in. This book is intended to aid the practitioners in meeting the specialized needs of their clientele.

There is a very important and exciting role for information professionals in the pursuit of furthering scientific knowledge for the benefit of humankind. This book is intended to act as an educational guide and resource for librarians, information specialists, and researchers interested in acquiring an understanding and developing a working knowledge of science and technology collections.

Our guide begins with an overview of the nature of scientific and technical literature, the information-seeking behavior of scientists and engineers, and a discussion of the research information cycle. We discuss how to get familiar with your collection; the myriad types of resources, materials, and publishers in the field; and how to assist your clients with their research.

Along with in-depth topical chapters we have provided an appendix that should be helpful to the library science student, the practicing professional, and the library science educator. The appendix includes 29 subject guides for the science and engineering fields.

Acknowledgments

Writing a book is a journey—the kind of travel that tests you along the way but entices you to continue around the bend just to see what's ahead. During our journey we were assisted by many colleagues and friends; we wish to thank:

Bo Baker (proofreader and image master);
Donna Beck (office mate extraordinaire and all-around encourager);
Andrew Bobick (expert assistance with subject bibliographies);
Denise Callihan (our patent expert);
Blanche Woolls (our adopted editor);
Colleagues and SLA members;
Current and past students from whom we have learned.

Chapter

Introduction to the Literature of Science and Technology

The published literature of science and technology is complex and enormous. In several aspects it is overwhelming, particularly when one considers extensive runs of printed journals. The journal literature alone has been published since 1665 and many books have even earlier imprint dates. No matter what the format, the basic function of scientific literature has been to serve as a foundation for advances in the various disciplines of science. As scientific advances and breakthroughs are recorded and made available as public record, each one adds to, refines, modifies, or refutes the existing record of scientific information.

Just as science is global, scientific literature is also global. The results of scientific research come from all areas of the world and the subsequent publications containing that research also come from all areas of the world, even though there may be differences in language, bibliographic style, and media delivery format. The literature of science records and documents the work of laboratory investigations that employ the scientific method.

Publication of experimental laboratory work accomplishes several things. In addition to reporting new scientific information, it permits the scrutiny and evaluation of the findings by others and allows comparison within a larger body of knowledge. Furthermore, other scientists whose ideas, methods, and contributions may be part of the investigation are generally acknowledged. Most important, the outcome of such investigations should be able to be replicated by anyone who wishes to validate the results of the research under consideration.

The journal literature in any given discipline is the public record of the research that has been undertaken, completed, published, and accepted by the scientific community. Journals and books represent two of the most common types of scientific literature. Other literature types range from patents and technical reports to standards and specifications. This literature is available in a variety of formats, ranging from print to online electronic resources. As new technologies continue to be developed and

1

improved, scientific and technical literature will be archived and disseminated in even more formats.

Scientific and technical literature can be divided into two major categories: (1) primary sources and primary publications of information, and (2) secondary sources of information. Primary sources and publications can be defined as those types that contain new and unique scientific and technical information. Information appearing for the first time in primary sources and primary publications of information has been generated as a direct result of research undertaken and completed in a laboratory. Examples of this type of literature include journals, conference proceedings, technical reports, patents, laboratory notebooks, and dissertations and theses.

Secondary sources are defined as those sources and publication types that compile, organize, analyze, synthesize, and repackage information from primary sources. There are many different types and formats of secondary information sources. Examples of this type of literature are almanacs, annuals, and yearbooks; bibliographies; biographical texts; dictionaries and thesauri; directories; electronic databases; encyclopedias; guides to the literature; handbooks, manuals, and tables; field guides and taxonomic literature; indexes and abstracts; manufacturers' literature and trade catalogs; monographs and treatises; reviews; standards and specifications; technical reports; and translations. One of the great benefits in consulting and using various secondary sources of information is in saving time, as they chronicle, compile, or describe specific facts, data, procedures, or related information from the primary literature. In searching other secondary sources such as indexes and abstracts and their electronic versions, an individual may locate material or information that was previously unknown to him or her. Within the scientific and technical community, both primary and secondary sources of information are required for conducting efficient and effective research. Sometimes precise information from a primary source (e.g., a journal) is required, while on other occasions, the time-saving feature of a secondary source (e.g., a handbook) is more important. Again, secondary publications are those that are derived from primary publications and subjected to various editing processes.

Primary information derived from research and development investigations in the laboratory can be communicated in a variety of channels prior to the publication of the final results in a scientific or technical journal. When the investigation is still in progress, there is much interaction among the members of the core research team as well as related groups or individuals within a particular funding grant or faculty. Sometimes this interest and interaction extend to members of the larger scientific community beyond a particular physical setting. At this stage, information flows in both directions: input on ideas and suggestions from the scientific community to the core research group and output that includes experimental data and preliminary findings from the search team to various groups. This interaction and communication are usually informal (including oral communication) but may include written notes and correspondence. Usually, there are no formal public records of such interactions and exchanges. The laboratory notebooks and other written records will, ultimately, form the basis for the formal primary publication(s) that may result in a technical report, a conference presentation, a dissertation, or a submitted journal article.

Another stage of the ongoing research may be some form of preliminary communication. When the research investigation has reached an advanced stage, the team may communicate the preliminary findings through a formal channel, perhaps as a letter or brief communication in a journal such as *Science* or *Nature*. Another option would be to submit a short article to one of the journals that specializes in publishing letters. This type of communication represents the earliest formal communication within the scientific literature. Such communications may provide new and breakthrough information on current research that may enable other investigators to stay abreast of current developments within a particular area. Short and preliminary communications are usually followed with a full-length paper that reports the complete and final research activities under investigation.

An additional possibility for presenting preliminary results is the presentation of a paper at a conference, which may result in one of several types of documents. The conference sponsoring body may distribute a preprint of the paper that will be presented. The presented paper may then be included in the published proceedings of the meeting, which may or may not be edited and include summaries of discussions and related material. Finally, a reprint of the presented paper may be distributed during or after the conference.

If the laboratory investigation is being done as part of the requirements for an advanced academic degree, the results are usually presented in the form of a master's thesis or doctoral dissertation. If the research was supported by an outside funding agency (e.g., the National Science Foundation or the National Institutes of Health), a report detailing and documenting research results may be required. Publications such as dissertations and theses as well as technical reports are an important form of the primary scientific literature and represent a substantial quantity of the total published output.

The final possibility for communicating the results of the completed laboratory investigation is the publication of the results in a refereed primary journal. This publication will result in a citation (i.e., author[s], title of article, title of journal that includes volume, date, and pages), which is the most important scientific bibliographic unit and constitutes the bulk of the primary literature of science.

While advances and breakthroughs occur in both science and technology, there are some fundamental differences between science and technology. Science is often referred to as "pure," while technology is "applied." For example, if an individual is studying the behavior of various metals under high temperatures, one could assume that this is science. Such laboratory investigations are usually performed by individuals who are referred to as scientists. When the research is focused on finding the ideal metal that can be incorporated into a lightbulb that will burn brighter, last longer, and be more energy efficient, the research can be called "applied." Applied research and related investigations are usually undertaken by engineers.

The literature that is most frequently reviewed and analyzed by scientists consists primarily of journals. Indexes and abstracts and/or electronic databases are gateways to productively searching the journal literature and retrieving relevant citations or the full document. Alternatively, engineers frequently use more technical types of literature, including handbooks and manuals, standards and specifications, manufacturers' catalogs, and trade journals.

In reviewing the development of science and technology, it appears that they have progressed unevenly. The Romans, for example, were technologists and made little contribution to pure science. Indeed, after the fall of Rome and until the Renaissance, pure science regressed. Before the Renaissance, science and technology developed independently, and one could have science without technology, and technology without science. In the Renaissance, however, science was reinvented. During the scientific revolution of the 17th century, science and technology began to connect, resulting in an important relationship. The melding of these two systems became permanent over the next two centuries and the interaction led to the acceleration in the progress of both. Indeed, modern science cannot be carried out effectively without technology, and modern technology cannot advance without science (Atalay 2004).

Added to the interaction between science and technology, another more recent alliance that has received attention is the relationship between the science establishment and the federal government. In 1945, the United States was thought to be not only the strongest economic and military power in the world but also the leader in science and technology. The development of a science establishment consisting of scientists, engineers, universities, industrial laboratories, and federal science agencies began early in the 20th century and has seen unprecedented growth that began during World War II and continues today (Mann 2000). In his book, Mann characterizes the relationship between science and the federal government (together with dates) as steps in a process that culminates in marriage. Specifically, he discusses: (1) love at first sight, 1939–1945; (2) courtship, 1945–1955; (3) marriage, 1955–1965; (4) end of the honeymoon, 1965–1975; (5) estrangement and reconciliation, 1975–1985; (6) golden anniversary, 1985–1995; and (7) the future, 2000 and later. Other monographs explore the growth and development of scientific policy and research as well as the ongoing relationship between the science establishment and the federal government (Foerstel 1993; Galison and Hevly 1992; Smith 1990; Westwick 2003).

REFERENCES

Atalay, Bulent. 2004. *Math and the* Mona Lisa*: The Art and Science of Leonardo da Vinci.* Washington, DC: Smithsonian Books.

Foerstel, Herbert N. 1993. *Secret Science: Federal Control of American Science and Technology.* Westport, CT: Praeger.

Galison, Peter, and Bruce Hevly. 1992. *Big Science: The Growth of Large-Scale Research.* Stanford, CA: Stanford University Press.

Mann, Alfred K. 2000. *For Better or for Worse: The Marriage of Science and Government in the United States.* New York: Columbia University Press.

Smith, Bruce L. R. 1990. *American Science Policy since World War II.* Washington, DC: Brookings Institution.

Westwick, Peter J. 2003. *The National Labs: Science in an American System, 1947–1974.* Cambridge, MA: Harvard University Press.

Scientific and Technical Communication

When scientists are engaged in research activities, they need to communicate regularly with other scientists. It is crucial that scientists report their work to others and to funding agencies; they need to know what research others are engaged in as well as taking part in discussions on topics of current interest to the scientific community. One-on-one discourse occurs in the workplace among colleagues, in electronic mail correspondence, through social networking tools on the Internet, and traditionally, by attendance at conferences and society meetings.

PROFESSIONAL ASSOCIATIONS AND SOCIETIES

Professional associations and societies are the key to communication and dissemination of knowledge and continuous discovery in science and technology. Professional societies bring together like-minded individuals to the benefit of all humanity.

Associations serve several purposes:

- Educating their members and the public
- Setting professional standards of conduct
- Setting and enforcing product safety and quality standard
- Encouraging volunteerism
- Informing the public
- Developing, compiling, and disseminating information and information policies
- Establishing forums for the exchange of ideas and information
- Making efforts to represent private interests
- Exercising political skills

According to a study funded by the American Society of Association Executives, 7 out of 10 Americans belong to one association; 1 out of 4 belongs to four or more associations. Associations offer education courses in their subject area as well as spend $14.5 billion on industry standard-setting activities each year. If you are a member of an association you may offer your services as a volunteer, both for association governance and the common good of the membership. Associations produce newsletters and journals and generally hold annual meetings for their members. The work of associations is often not understood by the general public and is not a visible activity to most people. Education is the most important benefit of membership, and in science and technology most associations are in the forefront of technological discoveries and advancements. Examples of associations for science and technology librarians are: the Special Libraries Association (SLA), the Medical Library Association (MLA), the American Society for Engineering Education (ASEE), and the Association of College and Research Libraries (ACRL), a division of the American Library Association (ALA).

Have you ever witnessed a court case that invited an expert to testify? Associations are often approached by government officials and legal counsel to provide the names of technical experts in key areas in order to inform a jury or to provide data in a court case, and may affect how a law or policy gets enacted. As an example, the American Medical Association's statistical data brought to light in governmental hearings can inform the decision-making body and assist in establishing important public policy on health issues that affect us all.

Associations range from purely social ones to scientific ones. Most are national, non-profit membership associations, but many in the sciences are international. The development of academies or societies was of great importance to science. First established in Europe, these societies were populated with small groups of men who met to discuss subjects of mutual interest. Financially, many societies were formed through the generosity of a wealthy member of society, a royal, or government official. Their enthusiasm and passion for science was the sole reason for forming this interest group. Through the strength of the society, members enjoyed freedom of expression and the free exchange of ideas.

HISTORICAL REVIEW

Prior to the development of the scientific journal, the major form of communication in the sciences was private correspondence. There were a number of drawbacks to this form of information exchange including: (1) a great deal of time and effort was required to write the letters, (2) letters were personal in tone and not sent to individuals who might disagree with the information presented, (3) some letters contained symbols, ciphers, and notations of shorthand that maintained secrecy, (4) many individuals who were interested in science did not receive the letters, and (5) issues of priority could not be satisfactorily resolved. Therefore, another form of communication was required.

The first scientific journal was published in France in 1665 as a result of a proposal that was first initiated that year. At about the same time, plans were developing in England to publish a scientific journal that contained the results of accounts of scientific research. This journal was to be published by the Royal Society of London and named the *Philosophical Transactions*. This new publication was issued on March 6, 1665, and consisted of 16 pages. Included in the initial issue was a dedication to the

Royal Society, an introduction, nine articles, and a listing of important philosophical books. The early publication record of the *Philosophical Transactions* was uneven due to events such as the plague in London and the great London fire. In spite of initial problems, the *Philosophical Transactions* has survived for almost three and a half centuries and is one of the most prestigious scientific journals still being published today. Some of the most important discoveries in science and related fields have appeared as papers in this journal. These initial two journals (*Le Journal des Sçavans* and the *Philosophical Transactions*) served as models for scientific periodicals that were issued by other European societies and academies.

The American Association for the Advancement of Science (AAAS) is the largest general scientific society in the United States. Founded in 1848 in Boston, it was mainly organized by geologists and naturalists. Furthering the work of scientists by helping facilitate cooperation, working to improve scientific methods in the promotion of human welfare, and increasing public knowledge and understanding of the role of science are its goals. AAAS is affiliated with nearly 300 scientific societies. Headquartered in Washington, DC, it publishes the weekly journal *Science.* In 1848 at the Great Hall of the Philadelphia Academy of Science, AAAS launched its tradition of migratory meetings to advance the course of science. In 1998, the association celebrated 150 years of advancing science. Several monographs have been published that center on the AAAS (Kohlstedt 1976; Kohlstedt and Sokal 1999; Kargon 1974; Wolfle 1989).

On March 3, 1863, President Abraham Lincoln signed a congressional charter creating the National Academy of Sciences in the United States. Today, the academy and its sister organizations: the National Academy of Engineering, established in 1964, and the Institute of Medicine, established in 1970, serve as the country's preeminent source of advice on science and technology and their bearing on national welfare.

The academies work through the National Research Council of the United States, established in 1916 at the request of President Woodrow Wilson to recruit science and technology specialists to participate and advise the government in times of peace as well as conflict. Today it is one of the world's most important advisory bodies. In a typical year, more than 6,000 scientists, engineers, industrialists, and health and other professionals comprise the National Research Council and participate on numerous committees. The council provides without compensation information and expert scientific advice to elected leaders, planning and policy makers, and the general public. Other activities of the council include meetings and workshops, consensus studies, program and research management, and educational activities. The National Research Council was reaffirmed and its charter broadened by President Dwight D. Eisenhower in 1956 and President George H. W. Bush in early 1993.

One of the oldest, international scientific societies is the Academy of Sciences' Russian Akademiya Nauk. Founded in St. Petersburg, Russia, in 1724, Russian Akademiya Nauk serves as the highest scientific society and principal coordinating body for research in the natural and social sciences, technology, and production in Russia. The academy, composed of outstanding scholars, is devoted to training students and publicizing scientific achievements and knowledge. Without the benefits of scientific societies, the achievements enjoyed by scientists worldwide would be hampered.

As information professionals working with sci-tech materials, it is important to note that associations and societies are frequently referred to by their name in acronym form. This naming convention can be puzzling at first when navigating through the maze of information sources in sci-tech. It is a common practice for a professional to refer to their association by its acronym, especially when conducting research in the library and seeking the proceedings or journal produced by that association. A few examples include IEEE (Institution of Electrical and Electronics Engineers), ACM (Association for Computing Machinery), SAE (Society of Automotive Engineers), ACS (American Chemical Society), and ASME (American Society of Mechanical Engineers).

THE RESEARCH INFORMATION CYCLE

How do scientists work? Generally speaking they follow a research pattern, or a cycle, that begins with the investigation of an idea or a topic. During this investigation period the researcher will look for information in a variety of places, utilizing both formal and informal communication paths. They will ask their fellow colleagues through personal and group contact. Whether through e-mail or during face-to-face interactions at a professional meeting, researchers inquire about their topic and may learn about past and current related experiments as well as research papers and their respective authors. Seeking out the leads they received on published papers takes them to the library. Here the formal method of investigation begins.

Searching within electronic databases and indexes, journal articles in that topic are discovered; perhaps a patent for an invention is acquired or a technical report is checked out. The library plays a clear role at this stage of the cycle. Information is extracted, reviewed, collected, scrutinized, and eventually compiled into a formal communication tool (a journal article, technical report, or presentation at a conference) and therefore disseminated back into the body of scientific knowledge offering a new and unique perspective on that topic. At this stage the findings are either accepted or rejected by the scientific community through peer review or are discounted as invalid research. As new knowledge becomes fact, it is communicated through textbooks and other reference sources. At this point in the cycle the information becomes obsolete as new research replaces it (Figure 2.1).

Stated another way, scientific methodology consists of first understanding or recognizing the problem, stating the problem, collecting data, analyzing that data, interpreting the findings, and drawing conclusions.

The Scientific Style Formula

Body of knowledge → Experiments → Axioms → Body of knowledge

Scientific research involves an interaction between tradition, experimentation, observation, and deduction. Scientists perform controlled experiments in a laboratory setting where they compare observations and measurements with a body of experimental data. Generally speaking, their field of expertise is very narrow and quite focused. Often their work is centered on a micro view of phenomena that adds to the larger body of knowledge in their area. Much of their work is performed in a laboratory setting.

Let's examine in more detail some of the important elements in the steps taken in the research cycle (Figure 2.1). *Informal communication* is a casual discussion or written note or memorandum that adheres less strictly to the rules and protocol. This state

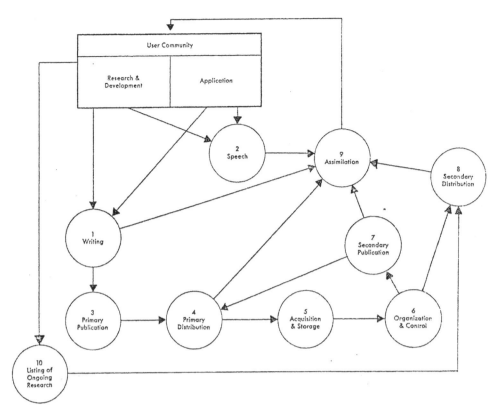

Figure 2.1. The research communications "cycle." From Lancaster, F. W., and Linda C. Smith. 1978. "Science, Scholarship, and the Communication of Knowledge." *Library Trends* 27(3): 368, figure 1. ©1978 by the Board of Trustees of the University of Illinois. Reprinted with permission of The John Hopkins University Press.

is most active in the early stages of research, the private phase, whereby the scientist is gathering information, discussing ideas with colleagues, conducting experiments, digesting findings, and preparing a manuscript for publication. It has been noted in numerous studies of informal communication that discussions and the exchange of preprints with other scientists working on related topics are critical steps in the conduct of effective research. As the amount of information in any one field can be prolific and difficult to monitor, it is essential to form a social organization with colleagues in their own field as well as in other fields that hold common research interests. This extended communication circle can be geographically diverse and reach beyond the boundaries of their normal research social circle (Crane 1969). Interdisciplinary research has become much more commonplace in the late 20th and early 21st centuries as communication technologies have enabled researchers to extend their communication patterns internationally.

An *invisible college* is a set of information communication relations among scholars or researchers who share a specific common interest or goal. Its first usage is credited to 17th-century members of the Royal Society of London, mathematicians who did not have a formal institution but called their affiliation to their group an "invisible college" based on their geographic proximity and regularly organized meetings called to discuss common scientific interests (Bartle 1995). To be a part of an invisible college requires an avid interest in expanding one's circle of intellectual curiosity beyond the traditional boundaries of the physical institution. It is a form of grassroots education, looking to and engaging in the larger community beyond normal circles of influence within a scholar's declared discipline and formal scientific field of study. It is a mode of informal communication.

Defining the invisible college, while seeming very straightforward, has changed since its first usage by Derek de Solla Price in his groundbreaking work (Price 1963) and has developed new direction enhanced by the development of new communication technologies and the changing information-seeking behaviors of researchers. A proposed new definition and research model (Zuccala 2006) has focused on three critical components: the *subject specialty,* the scientists as *social actors,* and the *information use environment.*

A *gatekeeper* is a key person who facilitates information transfer through the use of informal communication. Thomas Allen first suggested this concept while conducting groundbreaking studies in the flow of communication in the management of the research and development (R&D) process (Allen 1967). He suggests that the flow of technical knowledge can take several forms depending on the specialization of the gatekeeper and her or his information-seeking behavior. If a gatekeeper is a reader of the professional literature, this gatekeeper may benefit the laboratory by staying abreast of technological developments in the field. If a gatekeeper's information sources are mainly through contacts made external to the laboratory, the information could be helpful in providing specific details on particular research techniques. While both approaches invoke very different information-seeking styles, they can be of value to the organization to improve communication.

Formal communication is a presentation or document that strictly adheres to rules and protocol within the scientific community. Over three centuries ago the complex and controlled system that is adhered to today came into play, as the communication and resultant evaluation of scientific research changed from one of a private activity to a social activity involving recognition for the discovery and scientific contribution within the ranks of acknowledged scientists. To be the first to make an important and relevant scientific contribution is critical to establishing a successful career in science (Garvey 1979).

INFORMATION-SEEKING BEHAVIORS OF ENGINEERS AND SCIENTISTS

Information seeking is an important and ongoing part of the entire laboratory research cycle. Scientists are continuously seeking information, whether it is to be incorporated into the initial laboratory protocol or in the analysis and writing of the earliest results. Brown (1999) has investigated the information-seeking behavior of four groups of scientists including astronomers, chemists, mathematicians, and physicists. Anderson and colleagues (2001) studied how aerospace scientists and engineers used oral and written information sources while Ellis and Haugan (1997) explored the role of information in relation to research activities in different phases and types of projects in the industrial environment. Xu and colleagues (2006) examined in detail the various factors involved in information-seeking behavior focusing on the calculation of the seekers' cost-benefit analysis. In the academic environment, Tenopir and colleagues (2008) observed how users interacted with Elsevier's ScienceDirect information retrieval system in simulated class-related assignments.

Engineers

Engineering is a profession of devices, materials, systems, and structures. According to the Occupational Outlook Handbook, 2008–2009 edition: "Engineers apply the principles of science and mathematics to develop economical solutions to technical problems. Their work is the link between scientific discoveries and the commercial applications that meet societal and consumer needs." Most engineering training is conducted at four-year universities and colleges. An official body, the Accreditation Board, accredits these programs for Engineering and Technology (ABET) and is composed of practicing professional engineers. The field is very broad and includes the disciplines of aerospace; atmospheric and oceanic studies; agriculture; architecture; bioengineering; materials science; chemical, civil, computer and electrical, industrial, mechanical, nuclear, and petroleum studies; mining and geology; and naval architecture and marine research.

Upon graduation from an accredited engineering school an engineer is encouraged to work toward professional licensure. To be registered as a professional engineer one must take and pass the Fundamentals of Engineering Examination (FE/EIT), followed by acquiring work experience and subsequently passing the Principles and Practices of Engineering Examination (PE). Having the PE credential indicates that the engineer has met all the requirements of his or her state registration board to be able to gain a license to practice as a PE in that jurisdiction.

The field of engineering has hundreds of professional societies and tens of thousands of journals, newsletters, listservs, and related materials. Engineering resources are diverse and prolific. The half-life of engineering information is short, especially in the area of computers and electronics. Engineers value current information, but they have been reported to prefer to walk no farther than 90 feet to obtain it. Engineers favor electronic formats and there is widespread interest in learning more about tools for information retrieval. Engineers tend to be practical and make decisions based on data and evidence. They are often introverts by nature, are very focused, and appreciate when information is presented in an analytical and factual way. DuPreez (2007) has written an extensive review of engineers' information needs and their information-seeking behavior. In particular, components such as work roles, associated tasks, and outcomes are analyzed. Milewski (2007) examined the information-seeking strategies used by software engineers including the complex interaction of software developmental tasks and national cultural differences when working within a global team. In addition, this study investigated the preference of nonsocial sources when checking facts and documentation compared to social sources when diagnosing users' problem-solving questions. An investigation by Kraaijenbrink (2007) studied information obtained by engineers from the Internet and identified major information usage gaps.

Scientists and Engineers: How They Differ

The function of a scientist is to know, while that of the engineer is to do. Scientists look to nature and ask questions about how it operates, which generates more questions of interest. Scientists often work solo, whereas engineers tend to work in groups. The engineer does not have the luxury of solving problems that interest her or him but must solve problems as they arise, many times with costly social implications. The scientist can focus on a minute area of physical phenomena; while in the design process of engineering, the information needed to solve the problem can be more diverse and cross several fields. The scientist's final product is a published paper; the engineer's principal output is a product. See Figure 2.2 for an illustrated table on the information-processing elements in science and technology (Allen 1997).

Web-based electronic tools (search engines, databases, and full-text journals) have been the primary driver for the transition from paper-based research to electronic communications in the 21st century in the researcher's pursuit of scientific scholarly information. Academic scientists appear to be the early first adopters of electronic communication tools, as they need to communicate their ideas and findings rapidly. A census study of 902 academic scientists (Hemminger et al. 2007) captured their information-seeking behaviors, quantified the transition to electronic communications, and demonstrated how this affected different aspects of information seeking. Significant changes were found including increased reliance on Web-based tools, fewer visits to the library, and work being entirely done via electronic communication.

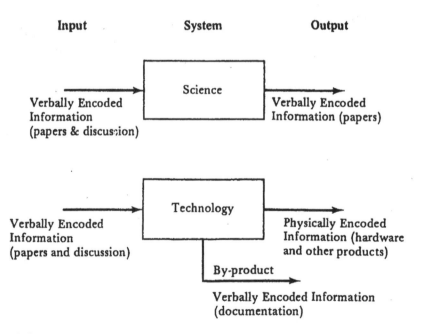

Figure 2.2. Information processing in science and technology. "Distinguishing Science from Technology." In *The Human Side of Managing Technological Innovation,* ed. Ralph Katz, figure 25.1 (1997). New York: Oxford University Press. By permission of Oxford University Press, Inc.

Fidel and Green (2004) studied the concept of the *accessibility of information sources* influencing the selection of information by engineers. They interviewed 32 engineers, who described their personal information-seeking behaviors in depth. Among factors cited as perceived by engineers were the right format and level of detail, a wealth of information in one resource, familiarity with the selected resource, and saving time in information gathering.

With an increasing use of the Web as a main vehicle for information seeking, it is important to build on past studies of information usage patterns commonly cited by engineers and scientists and build on that body of knowledge by analyzing in greater detail the challenges information-seeking researchers face as they deal with information overload and poorly designed Web sites and information tools.

REFERENCES

Allen, Thomas J. 1967. "Communications in the Research and Development Laboratory." *Technology Review* [MIT] (October/November): 31–37.

Allen, Thomas J. 1997. "Distinguishing Science from Technology." In *The Human Side of Managing Technological Innovation,* ed. Ralph Katz, 307–19. New York: Oxford University Press.

Anderson, C.J., M. Glassman, R.B. McAfee, and T. Pinelli. 2001. "An Investigation of Factors Affecting How Engineers and Scientists Seek Information." *Journal of Engineering and Technology Management* 18 (2): 131–55.

Bartle, R. G. 1995. "A Brief History of the Mathematical History." *Publishing Research Quarterly* 11: 3–9.

Brown, C. M. 1999. "Information Seeking Behavior of Scientists in the Electronic Information Age: Astronomers, Chemists, Mathematicians, and Physicists." *Journal of the American Society for Information Science* 50 (10): 929–43.

Crane, Diana. 1969. "Social Structure in a Group of Scientists: A Test of the 'Invisible College' Hypothesis." *American Sociological Review* 34 (3): 335–52.

DuPreez, Madely. 2007. "Information Needs and Information-Seeking Behavior of Engineers: A Systemic Review." *Mousaion* 25 (2): 72–94.

Ellis, D., and M. Haugan.1997. "Modelling the Information Seeking Patterns of Engineers and Research Scientists in an Industrial Environment." *Journal of Documentation* 53 (4): 384–403.

Fidel, R., and M. Green. 2004. "The Many Faces of Accessibility: Engineers' Perception of Information Sources." *Information Processing & Management* 40 (3): 563–81.

Garvey, William D. 1979. *Communication: The Essence of Science; Facilitating Information Exchange among Librarians, Scientists, Engineers and Students.* New York: Pergamon Press.

Hemminger, B.M., D. Lu, K.T.L. Vaughan, and S.J. Adams. 2007. "Information Seeking Behavior of Academic Scientists." *Journal of the American Society for Information Science and Technology* 58 (14): 2205–25.

Kargon, Robert Henry. 1974. *The Maturing of American Science: A Portrait of Science in Public Life Drawn from the Presidential Addresses of the American Association for the*

Advancement of Science, 1920–1970. Washington, DC: American Association for the Advancement of Science.

Kohlstedt, Sally Gregory. 1976. *The Formation of the American Scientific Community: The American Association for the Advancement of Science, 1848–1860*. Urbana: University of Illinois Press.

Kohlstedt, Sally Gregory, and Michael M. Sokal. 1999. *The Establishment of Science in America: 150 Years of the American Association for the Advancement of Science*. New Brunswick, NJ: Rutgers University Press.

Kraaijenbrink, J. 2007. "Engineers and the Web: An Analysis of Real Life Gaps in Information Usage." *Information Processing & Management* 43 (5): 1368–82.

Lancaster, F. W., and Linda C. Smith. 1978. "Science, Scholarship, and the Communication of Knowledge." *Library Trends* 27 (3): 367–88.

Milewski, A. E. 2007. "Global and Task Effects in Information-Seeking among Software Engineers." *Empirical Software Engineering* 12 (3): 311–26.

Occupational Outlook Handbook, 2008–2009 Edition, http://www.bls.gov/oco/ocos027.html.

Price, Derek J. de Solla. 1963. *Little Science, Big Science*. New York: Columbia University Press.

Tenopir, C., P. Wang, Y. Zhang, B. Simmons, and R. Pollard. 2008. "Academic Users' Interactions with Science Direct in Search Tasks: Affective and Cognitive Behaviors." *Information Processing & Management* 44 (1): 105–21.

Wolfle, Dael. 1989. *Renewing a Scientific Society: The American Association for the Advancement of Science from World War II to 1970*. Washington, DC: American Association for the Advancement of Science.

Xu, Y. J., C. Y. Tan, and L. Yang. 2006. "Who Will You Ask? An Empirical Study of Interpersonal Task Information Seeking." *Journal of the American Society for Information Science and Technology* 57 (12): 1666–77.

Zuccala, A. 2006. "Modeling the Invisible College." *Journal of the American Society for Information Science and Technology* 57 (2): 152–68.

3   Journals

The history of the scientific journal began in 1665. In that year, the French *Le Journal des Sçavans* and the *Philosophical Transactions* of the Royal Society of London started to publish the results of scientific research. Through this process, the journal was recognized as both a formal medium for disseminating scientific information and as a public record of the results and accomplishments of scientific investigations. The earliest scientific journals were the first vehicles in which investigators were able to report their findings. Today, the journal serves as a rapid messenger in disseminating scientific information. It also serves the important role of repository in which back issues of journals archive the accumulated knowledge over the span of years, decades, and even centuries.

The journal is a collection of original research contributions intended to communicate new research theories, report new phenomena, and disseminate ideas and experimental breakthroughs. It is felt by many that the journal is the most prestigious form of scientific and technical communication. Scientists need to have a vehicle in which to report their data in a timely way, and the journal serves this purpose to spread scientific discoveries. One might consider the number of scientists and their potential to submit scholarly articles for publication in scientific journals. In the early 1960s, it was stated that some 80–90 percent of all scientists who have ever lived were alive at that time (Price 1961). More recently it is the case that every doubling of the population has seen a tripling in the number of scientists, and today there are more working scientists than have previously existed in the entire history of science (Giddens 1999).

Journal articles generally follow a publication format:

- Abstract
- Introduction
- Outline of present research
- Results

- Conclusion
- References

The content of the journal is made up of articles or papers that include this standard publication format. The abstract is a summary of the major findings, together with a complete bibliographic citation of the author(s), title of article, title of journal, journal volume, part, year of publication, and pages.

The introduction to the paper discusses previous applicable research and states the objectives of the current investigation. The outline of present research covers the methods used to collect and analyze the data followed by the results, usually presented in graphs, tables, and photographs. The conclusion discusses the results and their significance together with a statement on further studies. Finally, a list of the literature cited throughout the paper is presented.

In addition to the journal articles or papers that contain these six standard sections, there are two shorter types of reports. One is known as a "note" or "brief paper" or "short paper." This type of report presents the results of parts of larger projects and investigations. It may also report preliminary results of ongoing research. A second type of shorter paper is known as "letters" or "communications." These are short papers on ideas that are new and unsubstantiated by extensive research. This type of paper presents results with a minimum of preliminary material. A more complete paper is expected to follow when the research has been completed.

TYPES OF JOURNALS

A number of different types of journals are currently being published. They include:

- Primary or archival journal
- Technical journal
- Trade journal
- Translation journal
- Popular science journal
- Review journal

Primary journals, also known as archival journals, are where original research in science and technology is published. This type of journal is often referred to as "scholarly." Journals of this type are held in the majority of science collections, especially academic and research institutions. Articles are extensive and in-depth and have been refereed by reputable scientists in the field. Due to the lengthy process of refereeing (also called peer review), publication is often delayed, but the prestige and protection of making sure that "good science" is reported is largely due to this time-intensive process. Examples of primary journals are:

- *Artificial Intelligence*
- *Biochemistry*
- *Journal of the American Chemical Society*

Technical journals are similar to primary journals and are published to meet the needs of the technical community, especially the technician in industry. Technical journals often contain short articles on practical applications of new techniques and examples in practice. Colorful and graphically attractive and including extensive advertisements, the information is presented in a style designed to appeal to the technical reader. Examples of technical journals are:

- *Iron Age*
- *Journal of Quality Technology*
- *Oil and Gas Journal*

Trade journals contain articles that center on recent patents, contracts, material prices, market trends, and related concepts. The articles are usually brief and emphasis is on customer-inquiry services and product evaluation. Examples of trade journals are:

- *Plant Engineering*
- *Tire Business*
- *Test & Measurement World*

Translation journals are those that are available in English but were originally published in a language such as Russian, Chinese, or Japanese. Translations of journals may be cover-to-cover, journals containing translation of individual articles chosen from different sources, or ad hoc translations of single articles. As you can imagine, it can be difficult to determine if a foreign-language journal has been translated; however, there are several places to determine and to locate information about the existence of a translated journal title.

In the mid-20th century a national translation center was established at the John Crerar Library in Chicago to assist users with not only locating journal titles that had translated versions but also providing the volume number and pages for these same articles in the translated version. Due to the nature of translating from one language to another, often the volume and page number changed once the translation was completed. This service was discontinued in the 1990s; however, these indexes to translations can still be found in some library collections as a finding aid.

Currently, major journals are produced in English, so the need for locating and translating articles has greatly diminished. If a translation is needed, however, there are many quality technical translation services available. A quick Web search or Yellow Pages search will assist in determining the right translation service for the job. A recent article discusses accurate and free Web sites for translation services (Martin and McHone-Chase 2009). Databases such as Chemical Abstracts can also be used as finding aids. Examples of translation journals are:

- *Journal of Organic Chemistry of the USSR* (translation of *Zhurnal Organichoeszoi Khimii* of the Academy of Sciences of the USSR)
- *Soviet Journal of Nuclear Physics* (translation of *Yadernaya Fizika*)
- *Astrophysics* (translated from Russian to English)

Popular science journals appeal to a large audience that includes the general public, the amateur, or the hobbyist. Written for the layperson, these publications are colorful, with lots of photographs, illustrations, and factoid boxes that explain "everyday science." News items are frequently featured and include elementary explanations of scientific phenomena. Examples of popular science journals are:

- *Current Science*
- *Popular Science*
- *Scientific American*

Journals can be either print or electronic; but the current environment favors electronic. It has become the norm for journals to be "born digital"; in other words, they are available only in virtual form. This is a popular method of dissemination and scholarly exchange for groups who need to communicate in a timely fashion because their fields, especially computer and information science, are changing rapidly. In general, most publishers are producing e-versions of their publications and the preferred purchasing method for journals in many academic environments is the electronic digital full-text method.

REVIEWS

Reviews are a unique form of scientific literature and one of the most important ways in which a scientist can keep up with scientific advances that are reported in the periodical literature. The number of scientific journal titles continues to grow substantially in each generation, and the number of scientific articles contained within these journals is even greater. It is becoming increasingly more difficult for an individual scientist to keep up with the relevant literature, even within a narrow area of research specialization. Reviews are one of the ways that scientists can cope with the seeming endless publication of journal articles, even within the narrow confines of one's research area. Essentially, reviews are surveys of the most important contributions and publications from the entire field of published scientific literature on a particular topic, for example, RNA polymerase. Reviews provide an overview, synthesis, and analysis of the subject since it was most recently covered in the literature. Reviews are important in dealing with the continuing proliferation of published scientific journal articles, the increasing specialization of these published articles, and the continued interdisciplinary nature of current scientific research. Reviews have been characterized as historical, biographical, descriptive, state-of-the-art, critical, and evaluative.

Furthermore, reviews are generally considered one of the more important forms of secondary scientific literature, that is, publications that do not report the results of original scientific research. However, reviews have unique and distinctive characteristics and some of the more important ones have been identified (Woodward 1974):

1. Original research is not reported in reviews.
2. Reviews usually appear in publications specifically devoted to publishing reviews (e.g., *Applied Mechanics Review*) or in a specialized section of periodicals reserved for reviews.
3. Reviews frequently have a title (from the journal table of contents) stating that the article is a review.
4. Reviews contain extensive references from the published journal literature.

The three major functions of reviews have been succinctly stated as: (1) current awareness, (2) tutorial, and (3) bibliographical (Subramanyam 1981). The current awareness function of reviews enables scientists to keep abreast of recent advances in a particular field of research. The tutorial function permits an individual to obtain knowledge in a particular subject outside a narrow area of specialization. Reviews are usually accompanied by extensive bibliographies of the cited literature and this feature fulfills the third function of reviews. One study showed that review articles are cited even more frequently than actual published journal articles (Garfield 1972).

Reviews may have titles such as:

- Annual Review of . . .
- Advances in . . .
- Critical Reviews in . . .
- Current Topics in . . .
- Progress in . . .
- Recent Advances in . . .
- Reviews of . . .
- Selected Topics in . . .
- Topics in . . .

JOURNAL PUBLISHERS

Journals originate from a variety of publishers including:

- Commercial firms
- Professional societies
- Universities
- Research institutes
- Government agencies

Commercial Firms

These are the for-profit publishers in science. Authors may have to pay to have their articles published in this type of journal, otherwise known as page charges. Commercial publishers are so efficient and well regarded that seeking publication in a commercial journal has become common practice for a researcher striving for a successful career in science. Elsevier is a large publishing house based in the Netherlands with corporate offices worldwide. Libraries pay very expensive subscription prices for these titles but can't exist and serve their clientele without these publications included in their collection.

Elsevier is part of Reed Elsevier Group, an internationally leading publisher and information provider. Operating in the science and medical, legal, education, and business-to-business sectors, Reed Elsevier provides high-quality and flexible information solutions to users, with increasing emphasis on the Internet as a means of delivery.

Professional Societies

As the promoter of science, societies publish the papers of its members and are usually far less costly than the commercial journal. Every field of science has multiple societies, so journals have become very specialized and prolific.

For example, the American Chemical Society (ACS) is a global society of 160,000 professionals in the chemical sciences worldwide. Membership includes the receipt of individual subscriptions to ACS journals and a free subscription to the weekly *Chemical & Engineering News* magazine.

Universities

Some universities, like the Massachusetts Institute of Technology (MIT) and Johns Hopkins, have presses of their own. They publish the work of their faculty, that of other universities and professional societies, and tend to fall in between the subscription price range of commercial publishers and professional societies.

The MIT Press is the only university press in the United States whose publication list is based in science and technology. This does not mean that science and engineering titles are the only ones they publish, but it does mean that they are committed to the edges and frontiers of the world and to exploring new fields and new modes of inquiry. The press publishes about 200 new books a year and over 40 journals. A major publishing presence in fields as diverse as architecture, social theory, economics, cognitive science, and computational science, they have a long-term commitment to "both design excellence and the efficient and creative use of new technologies." Their stated goal is: "to create books and journals that are challenging, creative, attractive, and yet affordable to individual readers."

Research Institutes

Many research institutes publish their own house organ. The Software Engineering Institute (SEI) at Carnegie Mellon University is funded by the U.S. Air Force and publishes its own journal, *News@SEI,* as well as technical reports and monographs.

Government Agencies

Many government-level agencies report their research projects via journals. Examples include the National Aeronautics and Space Administration (NASA), the National Institutes of Health (NIH), and the National Science Foundation (NSF).

PROLIFERATION OF THE JOURNAL

The number and growth of scientific and technical journals have been studied for almost as long as the journal has existed. David Kronick studied the early history of scientific and technical periodicals including those founded in the 18th century (Kronick 1976). The early 1800s were described as the "golden age of periodicals" that included "annuals, monthlies, and weekly-reviews, orthodox and heterodox-publications commercial, mechanical" (Subramanyam 1981, 33). In the early 1960s Derek de Solla Price postulated significant growth and an increase in the number of scientific and technical periodicals by a factor of 10 every 50 years beginning in 1700 (Price 1961). Three years later, a more realistic estimate by Price of the total number of scientific and technical periodicals actually being published was 30,000 (Price 1963). A more recent study, conducted by the British Library Lending Division, concluded that the number of available scientific and technical periodicals was 26,000 (Barr 1967). An extensive three-year project by King Research, Inc., indicated a total of almost 9,000 titles of scientific and technical periodicals in all fields of science published in the United States (King, McDonald, and Roderer 1981). Michael Mabe has studied the number of active, refereed scholarly, scientific, or learned journals and their rate of growth. While not limited to scientific and technical titles, the number of active titles is 14,694 (Mabe 2003). Mabe also identifies three recent, distinct time periods for journal development (1900 to 1940, 1945 to 1976, and 1977 to the present) and provides growth figures for each period (Mabe 2003).

The proliferation of scientific and technical journals continues, although the earlier prediction of 100,000 titles stated by Price (1963) seems unlikely. However, a more modest growth of titles continues and is due to such factors as:

- An increase in research and development activities
- An increase in the number of scientists and other personnel active in laboratory investigations
- The importance of the publication record as a measure a scientist's stature and reputation
- The increasing specialization and compartmentalization of science and technology

In addition to the proliferation of new journal titles, the "twigging" of established journals into smaller subspecialties is also ongoing. Credit for coining the term "twigging" is usually given to Curtis Benjamin, president of McGraw-Hill Publishing, who more than 35 years ago remarked that, as science grows, its tree of knowledge sprouts more and more twigs, but the size of the twigs doesn't change much (Swanson 2003). In twigging, the increasingly narrow specialization of scientists results in even more limited audiences and readers for publications. The increase in the number of journal titles is due not only to the birth of new journals but also to the "splitting" of existing journal titles. For example, at one time, the *Transactions of the American Society of Mechanical Engineers* was a single journal. Over time it has split and expanded into many other primary journal titles.

The large number of scientific journals provides many options for submitting manuscripts. Journal quality and reviewing standards vary widely, so there is almost always likely to be a place where research findings can be published. Within various scientific disciplines there are recognized rankings of journals and authors usually submit their manuscripts to the best journal possible. The problem of journal proliferation has been further compounded by the establishment of numerous scientific and technical journals begun by commercial publishers. In the "publish-or-perish" syndrome of research scientists and their employers, together with allegedly lower or less rigorous refereeing or screening standards, the quality of such commercial endeavors has been questioned.

Some commercial publications have very low standards and accept almost anything submitted in order to have something to print in the next published issue. In a study from 1972, it is found that scientific and technical journals published by commercial publishers were reported as costing 5 to 15 times as much as those published by societies (Moore 1972). More recent data confirms that this trend not only continues but has escalated. According to a recent study (Bergstrom 2010) in the field of neuroscience, the cost gap between the price per page of a for-profit journal vs. a non-profit journal is 11 percent; these costs are sharper still for the physics field, coming in at roughly a 30 percent difference from for-profit to non-profit.

Journal publishing has become increasingly expensive. Publication costs and subscription prices have increased steadily since the 1970s. To cope with the cost of publication, some publishers have established page charges, which were originally implemented by the American Institute of Physics (Barton 1963). Page charges are assessed by both society and commercial journals. In general, the page charges cover some of the costs of publication and allow the publisher to more fairly share charges between researchers and libraries. In some instances, support from page charges allows lower subscription prices and thus a greater circulation for the journal as well as open access after a certain period of time. Page charge forms are usually sent automatically on acceptance of an article for publication in a journal. A study commissioned by the Wellcome Trust (2004) to assess the actual costs of publishing scientific, technical, and medical research in peer-reviewed journals compared the costs between the current "subscriber-pays" model, where publishing services are free to authors and the article is published in a journal available via subscription, and an "author-pays" model where the author (or their funder or institution) pays for the publishing services but where the final paper is published in an open-access journal, available for free via the Internet to all who wish to use it. The report provides evidence that an author-pays model offers a viable alternative to subscription journals.

Refereed journals are publications whose articles or papers are reviewed by "expert readers" or "referees" prior to the publication of the material in a journal. The task of the referee is to make a critical evaluation of the submitted paper and a recommendation to accept or reject the paper for publication. In many cases, referees recommend acceptance (or re-evaluation) subject to a revision, and provide explicit guidelines for the author(s) to follow. The process of refereeing varies from one or two weeks to three or four months or longer. Referees are not supposed to communicate directly with the authors whose papers they are reviewing. All communication is through the editor. Refereed materials are also called "peer reviewed." Refereed or peer-reviewed articles are important and significant to the published scientific literature since they are, most importantly, scientifically accurate.

Having one's work refereed can be fraught with anxiety even for the most famous scientists. Such was the case for Einstein, who did not like to have his manuscripts reviewed. In the summer of 1936 he submitted a paper to the *Physical Review,* and a referee returned it with 10 pages of comments. Insulted, Einstein withdrew the paper so he could publish it elsewhere. He claimed the *Physical Review* had no right to show the paper to reviewers before publication, as was (and is) the American custom (Calaprice 1996).

Refereeing of journal articles is further compounded by the quantity of manuscripts received by an editor. In 1985, the newly appointed editor of *Science,* the academic journal of the American Association for the Advancement of Science (AAAS), discussed this problem. Daniel Koshland, who served as editor from 1985 to 1995, stated that: "For every 100 that we receive, we can publish no more than 20" and

> that all manuscripts will either be returned to their authors within approximately 2 weeks—so that the papers may be submitted elsewhere—or will continue in the reviewing process. Manuscripts that receive top ratings from the reviewing board will have a 50 percent chance of being accepted. Manuscripts will be evaluated in relation to others, not on the basis of an absolute standard. (Koshland 1985)

More recently, Peter Stern, editor at *Science,* was interviewed and asked about the number of manuscripts received and the submission process. In general, "it takes about ten weeks from submission to acceptance." Immediately after acceptance, the papers are posted on the Web and the copyedited print version follows within four to six weeks. Submission numbers "have exploded. About ten years ago, some 6,500 manuscripts were submitted . . . per year. In 2005, we received about 12,000, in 2006 13,000." Concerning acceptance rates, "in 2005, 8% of the submitted papers were published. Right now it's 7%." Overall, submissions appear to continue to increase, while the number of papers being published remains fairly constant (*B.I.F. FUTURA* Editors 2007).

The journal is a social institution that confers honor and reward on authors, editors, referees, and publishers. Published papers are a concrete measure of a scientist's contribution to the advancement of scientific knowledge. In addition, such papers are the basis for an evaluation of the laboratory work by peers and employers, particularly those that win grants, promotions, higher salaries, and tenure. In some academic environments there is significant ongoing pressure to publish. Editors and referees are critical to sustain and maintain the quality and integrity of the scientific literature. Publishers are rewarded with prestige and financial rewards.

In addition to the number of new journals that are being started, there are several areas that reflect the increased growth and development of an established journal title. One of these is, simply, an increase in the size of the journal, that is, an increase in the number of published pages due to lengthier manuscripts that probably reflect more sophisticated laboratory experimentation and data analysis. The increase in the size of the journal could also be a reflection of the number of additional papers that are submitted for review and ultimately accepted for publication.

Scientific journals escalated in cost and number of titles during the 20th century and this trend continues into the 21st century. Many journal titles have split off the main journal within a field to create sub-journals on highly specialized topics. Depending on a library's budget, it is almost impossible to consider purchasing every title published within a field, even some of the smaller, specialized ones. A common question is: How many science and technology journals are currently being published, and how quickly is that number growing? The number and growth characteristics of scholarly journals have been under debate for quite some time, so estimates vary depending on the actual number by the type of statistical model employed and the agency or research team attempting to answer the question. One such model has been compiled by two information scientists, Tenopir and King (2000), who are well-known researchers in the field

| Table 47 | Number of U.S. Scholarly Scientific Journals and Average Number of Issues, Articles, and Pages per Title by Nine Fields of Science: 1995 |

Field of Science	Number of Journals	Average Number per Title			
		Issues	Articles	Article Pages	All Pages
Physical sciences	432	14.2	306	2604	3342
Mathematics, statistics	206	9.1	127	2069	2276
Computer sciences	126	8.8	165	1947	2370
Environmental sciences	322	9.8	117	1641	1807
Engineering	828	9.0	163	1830	2039
Life sciences	2104	11.0	130	1396	1596
Psychology	342	4.5	49	757	842
Social sciences	2140	3.9	38	918	1099
Other fields/multi-fields	271	12.1	396	2742	4535
All fields	6771	8.3	123	1434	1723

Source: Tenopir and King 1997

| Table 48 | Number of U.S. Scholarly Scientific Journals and Average Number of Issues, Articles, and Pages per Title by Type of Publisher: 1995 |

Type of Publisher	Number of Journals	Average Number per Title			
		Issues	Articles	Article Pages	All Pages
Commercial	2679	9.9	118	1533	1811
Society	1557	9.3	202	1813	2296
Educational	1106	4.3	70	1500	1742
Other	1429	7.3	84	786	919
All publishers	6771	8.3	123	1434	1723

Source: Tenopir and King 1997

Figure 3.1. Number of U.S. journals published in 1995 by discipline and type of publisher. Tenopir, Carol, and Donald W. King. 2000. *Towards Electronic Journals,* 237. Washington, DC: Special Libraries Association.

of library science. They provide a comparison chart of the number of U.S. scholarly scientific journals published in nine fields of science for the year 1995 (Figure 3.1).

BUDGETING FOR JOURNALS

Every year journal costs increase. For the period 2000–2007, journal costs rose 9–12 percent every year. Each year EBSCO publishes *Serial Price Projections* as a service to libraries to assist with understanding and planning for the need to increase their budgets to handle this projected cost increase. EBSCO uses current price data received from publishers and historical price data to project price increases for the upcoming year. As always, it is prudent to be cautious in relying solely on price projections as the data is based primarily on historical trends and current estimates. Each library's budget considerations will be different and dependent on organizational funding. As an example, projections for 2008 for academic and academic medical libraries were set as falling 6–8 percent but with generally no such increase of the library's journal budget.

PRICING FACTORS

Many factors affect serials prices each year. Factors may include technology and electronic hosting costs for online titles, currency exchange rates, postage and handling costs related to print titles, changes to publisher pricing models, increases in pages and/or volumes, and overall general inflation. It is noteworthy that the U.S. Postal Service in the early 21st century increased their postal rates 11–17 percent by class of mail service. This rate increase significantly impacted the postage component of print journal prices, which was then passed on to the purchasing library's subscription. Finally, the takeover of the titles of one publisher by another can also significantly impact journal pricing.

Adding networking fees to the cost of the paper journal product, if the journal wishes to provide electronic access, requires increasing the budget line. Many publishers not only charge the cost of the original subscription but add a 10–50 percent fee to provide the electronic version. Additionally, if you wish to purchase the journal in electronic form, the price may go up hundreds of percent. A current example of this pricing model is the *Journal of the American Medical Association*. For three stand-alone workstations physically located inside a library building the cost is $525 per year. To offer the title on your library's Web site as an IP-authenticated journal (therefore not restricting the access to a single station physically located inside a library) raises the cost to $3,100, a cost increase of $2,575. Journal publishers have been expanding their options for product delivery formats, considered a plus when planning the library's strategy for information delivery to their clientele, but should also serve as a cautious note raising awareness of the potential for hidden costs such as annual licensing fees.

Serial subscriptions differ from book purchases in that they are not a onetime purchase but a long-term commitment and should be considered carefully during the budget planning process. Below is a set of tools and measures to consider utilizing when collection planning includes journal purchases.

Local Criteria (Depending on Type of Library) to Consider in the Selection of Journals

- Relationship to research and teaching
- Current or potential subscription request by clientele
- Relationship to other titles in the field
- Reputation of publisher/editor
- Type of publisher (society, commercial, university press, etc.)
- Cost
- Format (print, electronic, open access, etc.)
- Audience-appropriate level

Bibliographic Tools and Measures Used in the Selection of Journals

Ulrich's Periodical Directory/Ulrichsweb.com (ProQuest LLC): Known to information professionals as the authoritative source of bibliographic and publisher information. This directory covers more than 300,000 periodicals of all types: academic and scholarly journals, open access publications, peer-reviewed titles, popular magazines, newspapers, newsletters, and more from around the world. This tool is very helpful for determining where individual journal titles are indexed and, if purchased electronically, includes a powerful tool called the *Ulrich's Serials Analysis System.* This tool can assist when needing to identify, analyze, evaluate, and create reports about the library's serials holdings, aggregated journal collections, serials publishers, costs, and similar information (http://www.ulrichsweb.com/ulrichsweb/news.asp—serials).

OCLC WorldCat: WorldCat is a global network of library content and services utilizing the Web to search over 125 million bibliographic records that represent more than 1 billion individual items held by participating institutions. Traditionally a database that was used by catalogers and interlibrary loan librarians to learn of materials held in libraries internationally, WorldCat is now available to the public.

REVIEW SOURCES

Magazines for Libraries (R. R. Bowker): Learn from over 200 subject specialists working in the represented disciplines about serial titles of the most value when building your collection by browsing journal reviews in areas ranging from accounting to zines. Selections are based on titles that are most basic for the primary audience and by the expertise and experience of the selector. This tool will guide the user to the serials information required to build and maintain a quality collection that best meets the needs of the library's users. It reports essential information, useful statistics, and comparative data to assist in making collection decisions.

Association bulletins and newsletters: Library associations such as the American Library Association (ALA) and the Special Libraries Association (SLA) offer membership in subject divisions and produce bulletins that provide reviews of new journals in science and technology. Written by working sci-tech librarians, annotated listings provide bibliographic information as well as critical opinions on the value of the title.

Lists of recommended journals within disciplines: A quick search in a subject database or on an Internet browser will yield lists of recommended journal titles in all discipline fields in science and technology and are generally authored by information professionals or authorities in the respective fields/subfields of study.

Benchmarking of peer institutions: Everyone wants to understand how their institution measures against another institution, particularly those institutions most commonly considered a peer or competitor. Benchmarking is one approach that is a structured, proactive information-gathering tool that can assist an organization to develop comparative assessments of performance or practices across similar institutions. It is a process that sets an external standard against which internal operations can be measured, in this case, what journals are subscribed to, the analysis of usage against cost, and faculty publication records.

Interlibrary Loan (ILL) data: ILL activity is up in the United States (Beaubien 2007); Association of Research Libraries (ARL) statistics demonstrate that member libraries are borrowing more materials than they did 19 years ago. Online records of requests and statistical tools assist libraries in keeping track of individual journal titles requested. ILL data can inform collection decisions in the acquisition of future journal subscriptions based on local usage, disciplinary interest, and teaching and research needs.

Database/online availability: Research discovery has been dramatically improved with the advent of the World Wide Web and the increase in tools thereby extending a wider net in the quest for specialized searching. End-user tools such as databases, indexes, Internet browsers, and online full-text book services have put the power of knowledge acquisition more readily in the hands of the individual researcher.

Bibliometrics/citation analysis: Bibliometrics is the quantitative analysis of gross bibliographical units such as books and journal articles (Oxford English Dictionary 2008). Bibliometric studies provide a simplified picture of a complex set of variables and can be time-consuming and difficult. The application of mathematical and statistical qualitative and quantitative measures to the study of the usage of texts and information can yield information that can be used to inform collection management decisions. These measures are often applied to explore the impact of the published work of a set of researchers, a particular field/discipline, or even to measure the work of a single author. Commonly used to measure the growth and obsolescence of literature in subject fields, laws such as Bradford's Law (a small number of journals in a given field produce a significant number of relevant articles in that field) can be applied. *Citation analysis* is a collection management tool that helps the selector to determine what journals hold the most highly cited articles and can inform subscription decisions as well as cancellation decisions when managing a library journal collection within its budgetary constraints. This information can be of use when mapping the citation patterns of a new interdisciplinary field or to access the overall strength of a library's collection.

A published paper becomes part of the archival record of scientific publications. However, it may or may not be read. The only enduring record that it is read and used occurs as a citation in a later paper (Menard 1971). Some 90 percent of papers that have been published in academic journals are never cited, and as many as 50 percent of papers are never read by anyone other than their authors, referees, and journal editors (Meho 2007).

Specifically, *citation analysis* is the study of the frequency and patterns of citations in articles and books. Citations provide a measure of the quality of a publication and are analyzed for a variety of purposes ranging from the quality of an individual's publication record to the history and development of major scientific concepts by studying communication patterns of key individuals. Citation analysis uses citations in scholarly works, primarily books and journals, to establish links to other works or other researchers. Automated citation analysis has changed the nature of the research in this area, allowing millions of citations to be reviewed and analyzed.

Eugene Garfield, one of the founders of bibliometrics, started the Institute for Scientific Information (ISI) in 1955. Among his many innovative products are *Journal Citation Reports* (*JCR*). They provide information about academic journals in the sciences and social sciences and were originally published as a part of the Science Citation Index (SCI). A citation index is an index of citations between publications, allowing the user to easily establish which later documents cite earlier documents. This kind of data provides a basis for discovery of the publication history of any one author across his or her career and maps out related papers that impact findings of other publications. While the immediate use of this tool is to chart the path of scientific discovery and the key players behind those innovations, it is also used by academic and industry leaders to assess the publication record of their employees and has been known to be used as a method of evaluation in hiring, during promotion and tenure cases, and in the awarding of research funding. Caution should be noted here in judging scholarly effort based solely on this tool as not every journal or conference proceeding ever published is included in these databases.

Journal Citation Reports provide basic journal and citation information. The information given for each journal includes the basic bibliographic information consisting of publisher, title abbreviation, language, and ISSN. There are 6,166 journal titles in science and 1,768 titles in the social sciences. In addition, the subject categories for each journal are provided. The sciences have 171 categories and the social sciences have 54 categories.

The citation information includes basic citation data consisting of:

- The number of articles published during that year
- The number of times articles in the journal were cited during the year

Detailed tables show:

- The number of times the articles in the journal were cited during the year by later articles in itself and other journals
- The number of citations made from articles published in the journal that year to it and other specific individual journals during each of the most recent 10 years, including the 20 journals most cited
- The number of times articles published in the journal during each of the most recent 10 years were cited by individual specific journals during the year, including the 20 journals with the greatest number of citations

In addition, there is information on several measures derived from the data, including the *journal impact factor,* which is the ratio of the number of citations to the previous two years of the journal divided by the number of articles in those years.

The impact factor is, essentially, the average number of recent citations per article. It is frequently used as a proxy for the importance of a journal to its field. The *journal immediacy index* is also included, which is the number of citations that year to articles published the same year and reflects how quickly or rapidly an article is cited. Two other types of information are the *journal citing half-life,* which is the median age of the articles that were cited by the articles published in the journal that year, and the *journal cited half-life,* which is the median age of the articles in the journal that were cited by other journals during the year. For example, if a journal's half-life in 2005 is 5, that means the citations from 2001 to 2005 are half of all the citations from that journal in 2005, and the other half of the citations are before 2001.

Many commercial and society publishers of journals use citation data in advertising and other promotional literature. Figure 3.2 shows the use of citation data from

2008 ISI Data		
	Impact Factor	Total Cites
Optics Express	3.88	28,429
Optics Letters	3.772	37,689
Journal of Lightwave Technology	2.736	12,001
Journal of the Optical Society of America B	2.181	10,325
Journal of the Optical Society of America A	1.87	12,535
Applied Optics	1.763	31,492
	Impact Factor Ranking	Total Cite Ranking
Optics Express	3	4
Optics Letters	4	2
Journal of Lightwave Technology	8	7
Journal of the Optical Society of America B	11	9
Journal of the Optical Society of America A	16	8
Applied Optics	18	3

Figure 3.2. Citation data for the Optical Society of America journals. Reprinted with permission from The Optical Society, 2008 OSA Journals Catalog.

2008 for six journals published by the Optical Society of America (OSA). The data shows the impact factor, impact factor ranking, and the total number of citations for the six journal titles. The use of citation data is not a recent development in publicizing and promoting scientific journals.

As we have seen, citation analysis can be used as a collection development and management tool to determine which journals hold the most highly cited articles and allows the addition and cancellation of subscriptions based on concrete information when managing a journal collection within its budgetary constraints. In one classic study, it was determined that a relatively small core of 152 journals accounts for about half of all citations and that only 2,000 or so journals account for about 84 percent of all citations (Garfield 1972).

Although the Science Citation Index published by ISI has long been the most common tool for obtaining citation data, other services are now more readily available and challenging its dominance. Each service produces slightly different results, indicating the importance of using several citation sources to judge the importance and impact of a scientist's work. The Web is also leading to alternatives to the traditional *impact factor* of a journal or a particular scientist, including download counts and other data. In particular, Google Scholar has citation functionality and there is discussion about the possibility that this tool may have sufficient capabilities in the future to make commercial products unnecessary.

JOURNAL BUNDLING, ALSO KNOWN AS AGGREGATOR DATABASE

The term "journal bundling" refers to the practice of aggregating all titles produced by a publisher into a single product, or subject-based subsections, for sale as a complete and nonnegotiated product. This repackaged product is sold as a comprehensive unit that does not allow for the deletion or addition of the journal titles contained within. The purpose of an aggregator is to provide a single interface to widely dispersed, frequently changing information. The aggregation of a suite of journals owned by a single publisher has provided a new form of product delivery that clearly benefits the economic goals of the publisher. Marketed as an all-or-nothing deal, libraries are faced with of a loss of control in the selection of journal titles that best serve their community. The bundling of journal products has become the favored subscription model for the dominant commercial publishers of science, technology, and medical (STM) electronic journals, publishers such as Elsevier Science and Springer, to name just two.

Journal bundling practices have obvious benefits and disadvantages for libraries (Nabe 2001). On the positive side, being able to expand access to all the journal titles produced by a publishing house, rather than affording just select titles, strengthens the library collection. This effect is further enhanced if electronic resources are purchased via consortium deals, thereby further increasing the availability of a wide range of journals previously unattainable. So what is the downside of bundling? From a cost perspective bundling could become quite expensive as subscription prices rise and the bundle claims a larger percentage of the material budget to maintain. Flexibility in the selection of titles is not possible, leaving the library with an all-or-nothing decision process as cost-cutting measures become necessary. It may also prove that single journal titles from smaller publishing houses will be cut to be able to maintain the larger bundle deal, potentially creating a tension between funding larger bundled packages over the purchase of valuable and important single journal titles.

JSTOR: A HYBRID EXAMPLE OF JOURNAL SERVICE

JSTOR is a non-profit organization dedicated to helping the scholarly community discover, use, and build upon a wide range of intellectual content in a trusted digital archive (http://www.jstor.org). Working closely with the University of Michigan, JSTOR's founder, William G. Bowen, advanced his idea of converting printed scholarly journals to e-format, thereby having the capability to share the stored data in a centralized digital archive with libraries globally. Sharing electronic data could solve the challenge of the affordability, storage, and provision of the overwhelming product of research for research institutions. Promoting collaboration with organizations that could assist in achieving these objectives would maximize the benefits for the scholarly community as well as provide a stable, cost-effective, and discoverable archive for generations to come.

The organization holds over 1,000 academic journals in both digital and print formats representing work in the humanities, social sciences, and sciences. JSTOR has joined with other organizations to strengthen and broaden its collections to include other content types such as letters, oral histories, government documents, pamphlets, images, and 3-D models.

JSTOR's founding aim is to assist educational institutions by reducing costs associated with preserving and providing access to scholarship and other materials. Participation in JSTOR enables libraries to carry out de-accession or to move print holdings to shared off-site storage facilities, saving space for readers, reducing the need for interlibrary loan, and providing acquisition of complete digital collections at rates far less than they would have paid for print. A discussion of electronic journals would be incomplete without making ourselves aware of several movements in the library and scholarly research arena.

ELECTRONIC ACCESS VS. TRADITIONAL ACCESS

Scholarly publishing has been forever altered by the creation and adoption of the Internet and its global acceptance and practice as the preferred communication technology tool. Today in our collecting activities, we concern ourselves with the issue of access vs. ownership. If we buy digital copies of journals and lease them full text from publishers, do we really own them? When the lease runs out and we no longer can produce that article for a user or the computer system is temporarily down, did we do our job? These are critical questions for the science bibliographer to consider.

Researchers appreciate desktop delivery of full-text journal articles and are demanding more access every day. Purchasing journals electronically has become a cataloging challenge for libraries as we struggle to serve our patrons. Librarians have designed ways to catalog and point to issues that appear electronically for the patrons; and programs developed by vendors to assist users in locating full-text electronic versions of articles that their local library subscribes to are the key to seamless information delivery. The tool links from an article citation in a database to the full text of that article if it is a journal on the library's subscription list. As users search subject databases, if the library pays for access to the digital full-text version of an item located in their search, an icon is attached to that citation, providing a one-click access point.

The user simply clicks on that icon, and all options available for full-text delivery of that item appear on the user's desktop. The icon can be redesigned locally for ease of recognition by users and can include descriptive words such as "Get-It" or an institutions brand symbol. These software products work seamlessly for the user. Before the creation of tools that perform this function it was a time-intensive task for a researcher to determine the online availability of full text and to link to that version. Library staff works closely with the vendor to identify materials held in the print collection that correspond to electronic full-text copies. A service of this type can provide a high level of retrieval quality for users.

THE OPEN ACCESS MOVEMENT

Open access (OA) literature is digital, online, free of charge, and free of most copyright and licensing restrictions. The first open access journal, *New Horizons in Adult Education,* was launched in 1987 and gave rise to many experimental digital library projects throughout the 1990s, most notably the pre-Internet Gopher in 1991, the CERN preprint server in 1993, the creation of the NSF Digital Libraries Initiative in 1994, and Cite Seer in 1997. Digital collections were being spawned globally in the 1990s, and in 2001 a meeting was convened in Budapest by the Open Society Institute to discuss ways to accelerate progress globally toward making research articles in all disciplines freely available on the Internet. This landmark meeting explored ways to make open access effective and affordable for all. The attendees drafted the Budapest Open Access Initiative, a document outlining a statement of principle, a statement of strategy, and a statement of commitment.

Internationally, the lead supporter of the movement toward making research journals in science, technology, and medicine (STM) freely available to all for the common good of humanity was the United Kingdom's House of Commons Science and Technology Committee. After careful study the committee published a report in 2003–2004 outlining its findings and recommendations to the UK Parliament and detailing why the provision of scientific information was found unsatisfactory at that time. Its central thesis was that "due to a combination of publishers' pricing policy and the inadequacy of library budgets to meet the demands put on them by a system supporting an ever increasing volume of research, that change on all sides is necessary as a matter of urgency" (House of Commons 2003–2004).

Among the committee's recommendations was the creation of a UK interlinked network of institutional repositories for the deposit of all research articles produced in the country. This task had already started within SHERPA (Securing a Hybrid Environment for Research Preservation and Access), an existing consortium of 33 research organizations dedicated to building an open access repository across consortium partner institutions. Funding was recommended to support SHERPA to enable it to play a central role in this initiative as a fundamental first step to a major change in the way researchers disseminate their findings. Strong statements were made regarding the continuation of the peer review process in an open access agreement, to the practice that public libraries should have unlimited access to STM journals, and that government protection against the threat of closure to the British Library's Document Supply Service should be put in place. Recommendations also included closer study of the economics of publishers' business practices, such as journal bundling; the subscription-pays publishing model; and high publisher profit margins. Without reservation it was recommended that institutional repositories should be "a key component of any long-term strategy to ensure the preservation of digital publications" (House of Commons 2003–2004).

One nation could not lead this change alone; the UK government was encouraged to act as a proponent for change on the international stage and to lead by example. This example did indeed lead the United States to implement the NIH Public Access Policy, which states:

> The Director of the National Institutes of Health shall require that all investigators funded by the NIH submit or have submitted for them to the National Library of Medicine's PubMed Central an electronic version of their final, peer-reviewed manuscripts upon acceptance for publication, to be made publicly available no later than 12 months after the official date of publication: Provided, That the NIH shall implement the public access policy in a manner consistent with copyright law. (NIH 2008)

This policy ensures that the public will have access to the findings of NIH-funded research, both at home and abroad. An important milestone in the open access movement and the creation of institutional repositories was brought about in March 2008 by vote of the Faculty of Arts and Sciences (FAS) at Harvard University to mandate all FAS members to deposit their scholarly articles in an open access institutional repository, raising the bar for other U.S. universities (Albanese 2008). A key player and champion of the open access movement is Dr. Peter Suber, a research professor of philosophy at Earlham College, Indiana.

SPARC®, the Scholarly Publishing and Academic Resources Coalition, is an alliance of universities, research libraries, and organizations. The coalition was an initiative of the Association of Research Libraries (ARL) started in 1997 as a constructive response to market dysfunctions in the scholarly communication system. These dysfunctions have reduced dissemination of scholarship and crippled libraries. SPARC serves as a catalyst for action, helping to create systems that expand information dissemination and use in a networked digital environment while responding to the needs of academe. Leading academic organizations have endorsed SPARC (http://www.arl. org/sparc/about/index.html).

Furthermore, there is a detailed overview of the OA movement and continuing progress (http://www.earlham.edu/~peters/fos/overview.htm). Open Access is likely to benefit science by accelerating dissemination and uptake of research findings as reported by Eysenbach (2006) in his study on the citation reporting of open access articles. Several tools help information professionals when trying to assess whether or not a journal title falls under open access. The Directory of Open Access Journals lists journals already considered open access (http://www.doaj.org/articles/about).

INSTITUTIONAL REPOSITORIES (IR)

One cannot discuss the characteristics and benefits of creating institutional repositories (IR) without coupling the discussion with OA. The main organization behind OA is SPARC, an advocacy group developed by the ARL and whose members form an alliance of research and academic libraries seeking solutions to the uneven and costly scholarly publishing system.

Institutional repositories build on a growing grassroots faculty practice of posting research online, most often on personal Web sites but also on departmental sites or in disciplinary repositories. This demonstrates a desire for expanded exposure of, and access to, their work. In addition, digital publishing technologies, ever-expanding global networking, and enabling interoperability protocols and metadata standards are coalescing to provide practical technical solutions that can be implemented now. The convergence of these interrelated strands indicates that institutional repositories merit serious and immediate consideration from academic institutions and their constituent faculty, librarians, and administrators (Johnson 2002).

These institutional repositories include library collections, institutional archives, and databases (now online collections). It is also a service that offers a community a method for the management and dissemination of its members' scholarly digital works, as well as documents deemed to be of archival and scholarly significance. The creation of an IR involves the talent and support of information professionals across a library system, together with the technical expertise of computer professionals on the campus. One cannot be successful in building an IR without the buy-in of the larger stakeholder group, the campus community. More than an archive, a repository can be both a showcase that allows scholars to build scholarly profiles and a platform on which to publish original content in emerging open-access journals (Bankier and Perciali 2008). Aware of the underutilization of IRs at other institutions, researchers at the University of Colorado conducted a study that sought to understand the needs and goals of the community, because the university was planning an IR of its own. The results produced "personas" that described the different classes of future IR users on campus and helped to guide the planners in designing an IR that facilitated increased participation (Maness, Miaskiewicz, and Sumner 2008).

WHY AN IR?

- To make research freely available for the greater good of all; as mentioned previously, scholarly publishing is an uneven and costly system and with global acceptance of Web technology, it presents an opportunity to responsibly share the world's wealth of information.
- To make research findings easy to access (electronically available at institutional sites and searchable at the browser level by internal and external community members); by the act of self-archiving, an author's work receives wider dissemination (Harnad et al. 2008).
- To create cross-disciplinary research and to facilitate the building of global virtual subject networks.
- To raise issues of authors' rights.

In addition, IRs are a way for scholars to control research space on their terms, outside of the traditional means of the publication of their work through commercial or professional society journals. Self-archived work is called "grey" literature; it is hard for others to find, to use, and to share in a timely manner and, therefore, a formalized repository eliminates these concerns. The NIH in the U.S. now requires grant-funded authors to make their work publicly available within 12 months after traditional publication. The deposit of said work in PubMed opens up research to all. PubMed is a service of the U.S. National Library of Medicine that includes over 18 million citations from MEDLINE and other life science journals for biomedical articles back to 1948. PubMed includes links to full-text articles and other related resources.

An institution must be prepared to commit resources when considering creating an IR on campus. Funding will be required to hire or assign existing staff to the planning and management of the repository; to select and purchase a software platform; to market, implement, and populate the site with documents; and to growth it. There are many benefits to the implementation of an institutional repository including the ownership, regulation, and unique goal setting of the archive of the intellectual product created by the community it represents, therefore becoming the ultimate indicator of an institution's academic quality (Nath et al. 2008). Perpetual and cumulative, its contents will be preserved, discovered, and used by generations to follow. According to the Registry of Open Access Repositories (ROAR), there were 1,239 active organizational depositories globally as of late 2008 (http://roar.eprints.org/index.php) (see Figure 3.3).

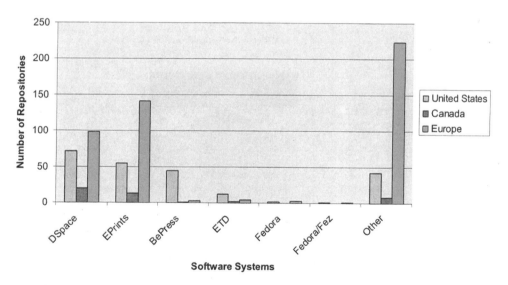

* Data from ROAR, March 3, 2008.

Figure 3.3. Repository software in the United States, Canada, and Europe. Campbell-Meier, Jennifer. 2008. "Case Studies on Institutional Repository Development: Creating Narratives for Project Management and Assessment," 118. PhD diss., University of Hawaii.

As traditional publishing practices change and more documents are "born digital," IRs are being studied by scholars as a means to collect digital-only information, as well as to provide access effectively and at low cost to the parent organization and to global scholarly audiences. One such study by Campbell-Meier investigates the development of repositories at doctoral institutions, reporting on influences in their development across the six institutions chosen for the study, as well as capturing best practices and providing cross comparisons in the adoption of an IR. Notably the use of "storytelling," creating a narrative for IR development, is discussed (Campbell-Meier 2008) (see Figure 3.4). As sketched in Figure 3.5, Paul Walk illustrates his vision for discovering scholarly information.

Case Study Site	Perceived Benefits
Indiana University	• Scholarly communication discussions between faculty and librarians • Content accessible to researchers • Providing a home for undergraduate research
Indiana University Purdue University at Indianapolis	• Interacting with faculty • Being involved in something new to the library world • Centralized space for particular content – e.g., Conference proceedings
Purdue University	• Access • Collects the intellectual output of the institution • Value adding to content (not just a storage spot) • Problem solving (id 373) • Working with faculty and departments in new ways • Making existing resources more usable/accessible
Simon Fraser University	• Experience with digitization projects • Central space to store items • Creating a "permanent" archive (ID 297) • Provides a "safe" home for digital materials • Visibility for authors • Visibility for the institution • Providing open access to materials
University of British Columbia	• A chance to make discipline specific material more accessible • Being able to "put stuff out there in a systematic way" • Make the intellectual output of the university accessible • Creating a leadership role for the librarian in sc • Support the learning, research, teaching of a university • Showcase for work that happens at a university • Building bridges across campus (faculty, IT services...) • Taking the library in a new directions • Becoming more involved with research on campus
University of Washington	• Identify the scholarly output of the institution • Store and preserve hybrid and digital content • Central system • Provides a "space" for and access to collections that would not be easily accessible elsewhere • Aids dialogue about abstract concepts

Figure 3.4. Case study sites and perceived benefits. Campbell-Meier, Jennifer. 2008. "Case Studies on Institutional Repository Development: Creating Narratives for Project Management and Assessment," 141. PhD diss., University of Hawaii.

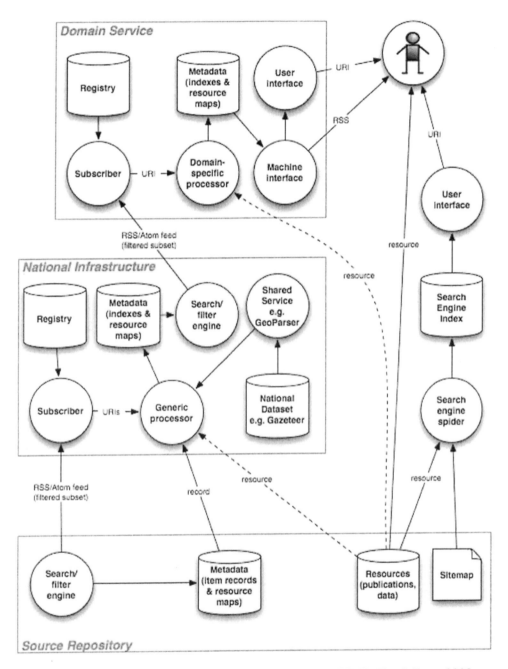

Figure 3.5. Diagram 2: Discovery of scholarly work. Walk, Paul. June 2008: http://blog.paulwalk.net/tag/ukoln/.

In this chapter we explored the importance of journal literature to science and technology disciplines and the latest trends in making this information more affordable and therefore widely available to all. In chapter 5 we will discuss ways to stay informed about new materials and innovations in the STM fields and how to alert our clients to the new publications, a practice known as current awareness services.

REFERENCES

Albanese, Andrew. 2008. "Harvard Mandates Open Access." *Library Journal* 133 (5): 16–17.

Bankier, Jean-Gabriel, and Irene Perciali. 2008. "The Institutional Repository Rediscovered: What Can a University Do for Open Access Publishing?" *Serials Review* 34 (1): 21–26.

Barr, K.P. 1967. "Estimates of the Number of Currently Available Scientific and Technical Periodicals." *Journal of Documentation* 23 (2): 110–16.

Barton, H. A. 1963. "The Publication Charge Plan in Physics Journals." *Physics Today.* 16 (6): 45–57.

Beaubien, A. 2007. "ARL White Paper on Interlibrary Loan." Association of Research Libraries. Available at: www.arl.org/bm%7Edoc/ARL_white_paper_ILL_june07.pdf.

Bergstrom, Ted. 2010. *The Office of Scholarly Communication: The Economics of Publishing.* Santa Barbara, CA: UC Office of Scholarly Communication. Available at: http://osc.uni versityofcalifornia.edu/facts/econ_of_publishing.html.

B.I.F. FUTURA Editors. 2007. "Gatekeepers of Science: Interview with Peter Stern, Editor at *Science* Magazine." *B.I.F. FUTURA* 22 (1): 14–18.

Calaprice, Alice, ed. 1996. *The Quotable Einstein.* Princeton, NJ: Princeton University Press.

Campbell-Meier, Jennifer. 2008. "Case Studies on Institutional Repository Development: Creating Narratives for Project Management and Assessment." PhD diss., University of Hawaii.

Eysenbach, G. 2006. "Citation Advantage of Open Access Articles." *PLoS Biology* 4 (5): e157. DOI:10.1371/journal.pbio.0040157.

Garfield, Eugene. 1972. "Citation Analysis as a Tool in Journal Evaluation." *Science* 178 (4060): 471–79.

Giddens, A. 1999. *Runaway World: How Globalisation Is Reshaping Our Lives.* London: Profile Books.

Harnad, S., et al. 2008. "The Access/Impact Problem and the Green and Gold Roads to Open Access: An Update." *Serials Review* 34 (1): 36–40.

House of Commons Science and Technology Committee. 2003–2004. *Scientific Publications: Free for all?* Available at: http://www.publications.parliament.uk/pa/cm200304/cm select/cmstech/399/399.pdf.

Johnson, Richard K. 2002. "Institutional Repositories: Partnering with Faculty to Enhance Scholarly Communication." *D-Lib Magazine* 8 (11). DOI: 10.1045/november2002-johnson.

King, Donald W., Dennis. D. McDonald, and Nancy K. Roderer. 1981. *Scientific Journals in the United States: Their Production, Use and Economics.* Stroudsburg, PA: Hutchinson Ross.

Koshland, Daniel E. 1985. "An Editor's Quest.2." *Science* 227 (4684): 249.

Kronick, David. A. 1976. *A History of Scientific and Technical Periodicals: The Origins and Development of the Scientific and Technical Press, 1665–1970,* 2nd ed. Metuchen, NJ: Scarecrow Press.

Mabe, Michael. 2003. "The Growth and Number of Journals." *Serials* 16 (2): 191–97.

Maness, J. M., Tomasz Miaskiewicz, and Tamara Sumner. 2008. "Using Personas to Understand the Needs and Goals of Institutional Repository Users." *D-Lib Magazine* 14 (9/10). DOI:10.1045/september2008-maness.

Martin, Rebecca A., and Sarah McHone-Chase. 2009. "Translation Resources on the Web." *C&RL News* 70 (6): 356–59.

Meho, Lokman I. 2007. "The Rise and Rise of Citation Analysis." *Physics World* 20 (1): 32–36.

Menard, Henry W. 1971. *Science: Growth and Change.* Cambridge, MA: Harvard University Press.

Moore, James A. 1972. "An Inquiry on New Forms of Primary Publications." *Journal of Chemical Documentation* 12 (2): 75–78. DOI: 10.1021/c160045a002.

Nabe, Jonathan. 2001. "E-Journal Bundling and Its Impact on Academic Libraries: Some Early Results." *Issues in Science & Technology Librarianship* 30. Available at: http://www.library.ucsb.edu/istl/01-spring/article3.html.

Nath, S. S., et al. 2008. "Intellectual Property Rights: Issues for Creation of Institutional Repository." *DESIDOC Journal of Library & Information Technology* 28 (5): 49–55.

National Institutes of Health (NIH). 2008. NIH Public Access Policy Implements Division G, Title II, Section 218 of PL 110–161 (Consolidated Appropriations Act).

Price, Derek J. de Solla. 1961. *Science since Babylon.* New Haven, CT: Yale University Press.

Price, Derek J. de Solla. 1963. *Little Science, Big Science.* New York: Columbia University Press.

Subramanyam, Krishna. 1981. *Scientific and Technical Information Resources.* New York: Marcel Dekker.

Swanson, Don R. 2003. *A Literature-Based Approach to Scientific Discovery.* Presentation at UIC September 4. Available at: http://128.248.65.210/cci/workshop/swanson.ppt.

Tenopir, Carol, and Donald W. King. 2000. *Towards Electronic Journals.* Washington, DC: Special Libraries Association.

Wellcome Trust. 2004. *Costs and Business Models in Scientific Research Publishing.* Report commissioned by the Wellcome Trust. Compiled by SQW Ltd. Available at: http://www.wellcome.ac.uk/About-us/Publications/Reports/Biomedical-science/WTD003185.htm.

Woodward, A. M. 1974. "Review Literature: Characteristics, Sources, and Output in 1972." *ASLIB Proceedings* 26 (9): 367–76. (This work is licensed under a Creative Commons Attribution 2.0 UK: England & Wales License.)

OPEN ACCESS WEB SITES

Directory of Open Access Journals: http://www.doaj.org/

Directory of Open Access Repositories: http://www.opendoar.org/

Open Access News: http://www.earlham.edu/~peters/fos/fosblog.html

Public Knowledge Project: http://pkp.sfu.ca/?q=ojs

SPARC: Scholarly Publishing and Academic Resources Coalition: http://www.arl.org/sparc/about/index.shtml

INSTITUTIONAL REPOSITORY WEB SITES

DSpace: http://www.dspace.org/

EPrints: http://www.eprints.org/

Registry of Open Access Repositories (ROAR): http://roar.eprints.org/index.php

Selected Works on Institutional Repositories: http://works.bepress.com/ir_research

SHERPA: http://www.sherpa.ac.uk/

Chapter

Specialized Databases

Scientific publishing in the late twentieth century was so prolific that it produced millions of written documents recording science's advancement and knowledge. In fact, over 95 percent of all scientific breakthroughs in human history happened in the decade of the 1990s. With so much information out there to organize and find, we are delighted that modern computer technology has provided the solution and the means for locating published documents. The technology of online databases has transformed our ability as information providers to serve our user community. It has changed the face of our facilities and the way we do business.

What is a database? It is a computer-readable electronic library. Databases are generally subject specific and contain citations for books, journal articles, proceeding articles, and other documents. Many databases are organized and made available to librarians via a commercial vendor but today many organizations are scanning documents locally and creating their own in-house databases. Many of these local databases contain information known as "documents" or "information" created in electronic digital form, and not the one digitized through scanning (BusinessDictionary.com 2009).

Commercial databases were introduced to libraries in the 1970s and required the information professional to learn a search language to manipulate the data and to conduct the search for the patron. Databases may contain a mixture of citations to documents with some full-text documents or provide the full text of documents. Some databases are text-based while others contain graphics, statistical data, or figures and enable the drawing of chemical formulas. Most specialized databases in sci-tech are fee-based while some are freely available on the Internet.

An example of a free quality database is The National Library of the Environment (NLE) produced by the National Council for Science and the Environment (NCSE, see http://ncseonline.org). Figures 4.1 and 4.2 show samples of NLE search screens. This database is government sponsored, publicly accessible, and can be considered a prototype for information delivery in the future.

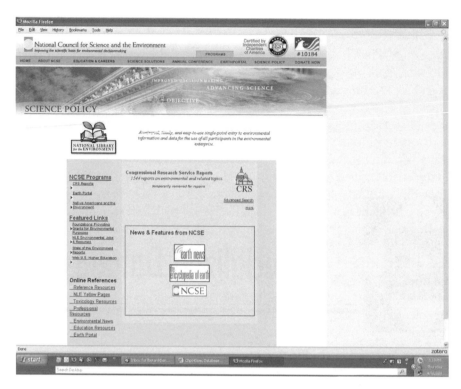

Figure 4.1. NLE homepage.

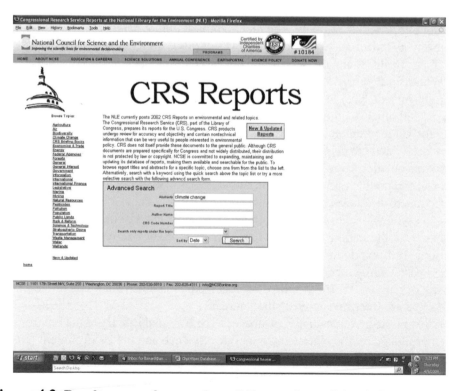

Figure 4.2. Results screen for search on "climate change" in all fields, NLE database. Accessed April 2009.

An example of a free "search engine/database" is Google Scholar (http://scholar.google.com/). Google Scholar acts as a gateway for searching across many disciplines and for locating scholarly literature resources in one location. Sources located here include peer-reviewed papers, theses, books, abstracts, and articles from academic publishers; professional societies; preprint repositories; universities; and other scholarly organizations. Once a scholar selects their institutional library from Google Scholar's list, the program can then determine which journals and papers the home institution is subscribed to electronically. For each item returned in the search list, a link back to the source is provided (see Figure 4.3).

To be proficient in any field, a researcher must have access to databases. Journal article citations or the actual full text of journal articles are the main type of document found in technical databases. Depending on the database interface, a variety of searches can be performed. Typical search fields include author name, title, keyword, journal title, subject or descriptor, and an abstract of the article. Most search engines provide a novice and an expert search screen, with options to sort the search results by date, record type, and other limits. See Figure 4.4 for the elements of a typical database record. If you have not had much experience searching, you may wish to check out the *Manual of Online Search Strategies,* Sciences volume (Armstrong, Large, and Large 2000). It will assist you in learning about engineering and science databases

Figure 4.3. Screen print of a typical Google Scholar search for the phrase "cell signal*." Accessed February 2009.

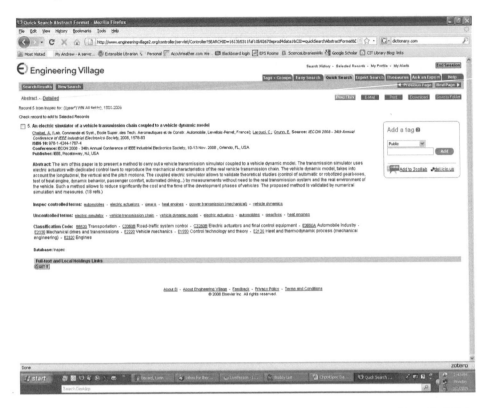

Figure 4.4. The elements of a typical database record. Accessed March 2009. Copyright Elsevier Inc.

and in formulating search queries. The literature has many tutorials and guides that address general search techniques such as *Librarian's Guide to Online Searching: 2nd edition* (Bell 2009) and *Reference and Information Services in the 21st Century: An Introduction* (Cassell and Hiremath 2006).

THE DATABASE SELECTION PROCESS

Hundreds of databases are available, mainly for a fee. The holdings of libraries vary widely and in essence depend on the size of the budget when deciding on purchasing online materials. Making the choice to purchase electronic products involves much more than just finding the money; it also involves purchasing equipment, making space for the equipment, allocating or purchasing the ergonomic furniture to house the equipment, the handling of maintenance contracts, and the provision of local technical support services. Training for staff is also a budgeted item to consider. Legal factors include understanding a technical license and being prepared to live within the constraints of what clients are allowed to access under the license agreement.

The first consideration when purchasing databases is your audience. You need to know the disciplines and subdisciplines within those fields that are of interest to your audience and be prepared to evaluate databases you should purchase to satisfy those information needs. You will be working with commercial vendors and informing yourself about products they offer and on what platform the information is available. Many versions of the same product can be purchased from several vendors with different interfaces, offering the librarian real choices for platform access and cost. An example of a database made available to purchasers by several vendors is Chemical Abstracts.

Chemical Abstracts is a database in the chemistry field that indexes over 30,000 journal titles and books, contains over 30 million chemical compounds, and is adding pre-1907 records to the database. You can purchase access to it from several vendors, including STN and DIALOG as well as from the parent company, Chemical Abstracts Service (CAS). The content is basically the same but the search interface design varies by company. In the case of the method of delivery from the parent company, a variety of options abound. A researcher (such as a graduate student or lab technician) may find SciFinder Scholar to be the best interface for specific chemical information. SciFinder Scholar contains chemical patents, references to scholarly journal articles, and includes access to the CAS Registry database, the original source and final authority for CAS Registry Numbers. The researcher may also opt to access CAS information via the STN system, especially in the quest for a broad range of sci-tech literature. A patent specialist or information professional will find that STN can be mined in a variety of ways and in the case of patent information, access to the Science IP Search Service reports will locate in-depth information on prior art/patentability for the prosecution of new patents, freedom to practice/operate for new and existing technologies, and patent validity for infringement litigation. Each of these search services are available via the Internet, as software loaded on a local area network or in CD-ROM format. As of January 2010, the traditional print copy is no longer being produced. Whatever the local condition can support, CAS offers a method of delivery appropriate to the audience.

Pricing factors will vary by vendor. In an academic library it is common for the vendor to charge by the number of full-time equivalent (FTE) students enrolled at the institution, by the available "seats" or "concurrent users" utilizing the database at the same time, or by password. Oftentimes information is "bundled," a concept we covered in chapter 3 in our discussion on journals, and it is important to request details on what titles are included in the bundled package. Be sure to ask the vendor you have chosen if they have any clauses in their contract for "digital rights management" (DRM) and thereby impose limitations on the usage of digital content and devices, such as charging for printing or downloading, or restricting the number of downloads permitted. If such restrictions apply, this will have an impact on your ability to provide access to information to your users.

Functionality and database features need to be determined based on your users' information-seeking behavior and needs. Allison, McNeil, and Swanson (2000) offer a list of interface features and design to consider when selecting delivery options as well as searching options and cost and content considerations.

What are some very important databases in sci-tech? Every library situation varies. Below is a short list of critical ones to consider purchasing for an engineering or science library:

1. Science Citation Index (also known as ISI or Web of Science)
2. Chemical Abstracts (SciFinder Scholar)
3. Inspec (formerly Computer & Control Abstracts)

    4.   NTIS (National Technical Information Service)
    5.   Engineering Index (in e-form called EI Compendex)
    6.   Applied Science & Technology Abstracts
    7.   Metadex (Materials Science Collection)
    8.   IEEE Explore (electrical engineering and computer science literature)
    9.   Environmental Science and Pollution Abstracts
  10.   Scopus

The list above represents just a few of the main ones. There are so many databases to choose from, and depending on the field of interest, some very specialized ones are available, such as:

1. Biological Abstracts
2. CINAHL
3. GeoRef
4. MathSciNet
5. Medline
6. PsychInfo
7. PubScience
8. Dissertation Abstracts

Not every institution can purchase all the pertinent databases produced for each field and purchasing decisions will be tailored to local programs and research need. A solid, basic engineering database with a tradition of service since 1884 is the Engineering Index (online known as EI Compendex). This database is a comprehensive interdisciplinary engineering database with abstracts from over 5,000 international journals, conference papers, and technical reports and is the ideal starting place for any engineering research topic search. It covers all the major disciplines of engineering and has continued to do so for over 100 years. Subjects include civil, energy, environmental, geological, and biological engineering; electrical, electronics, and control engineering; chemical, mining, metals, and fuel engineering; mechanical, automotive, nuclear, and aerospace engineering; computers, robotics, and industrial robots. One of the great strengths of EI Compendex is its depth of subject indexing. See Figure 4.5 for a view of the main search screen.

The Science Citation Index (SCI) resource enables a scientific community to gain access to historical research and to keep abreast of recent developments in their respective disciplines. One of the pluses in purchasing this database for a library is its interdisciplinary nature, as it contains subject databases covering the sciences, social sciences, as well as the arts and humanities. A single platform extends and deepens research coverage by integrating journal, patent, proceedings, and life science literature with Web resources and other scholarly content.

As can be viewed in Figure 4.6 there are a variety of ways to conduct a general search. One can search by topic, author, publication name, language, or document type, to name a few criteria. Ways to limit your search can include the time span of your search (one year or all years covered by the database) and a choice of databases your topic falls within (pure science, social science, or humanities and the arts). A history of your search strategies are saved under "Search History" and can be used to formulate additional searches. You can also elect to "mark" those results of most interest to your

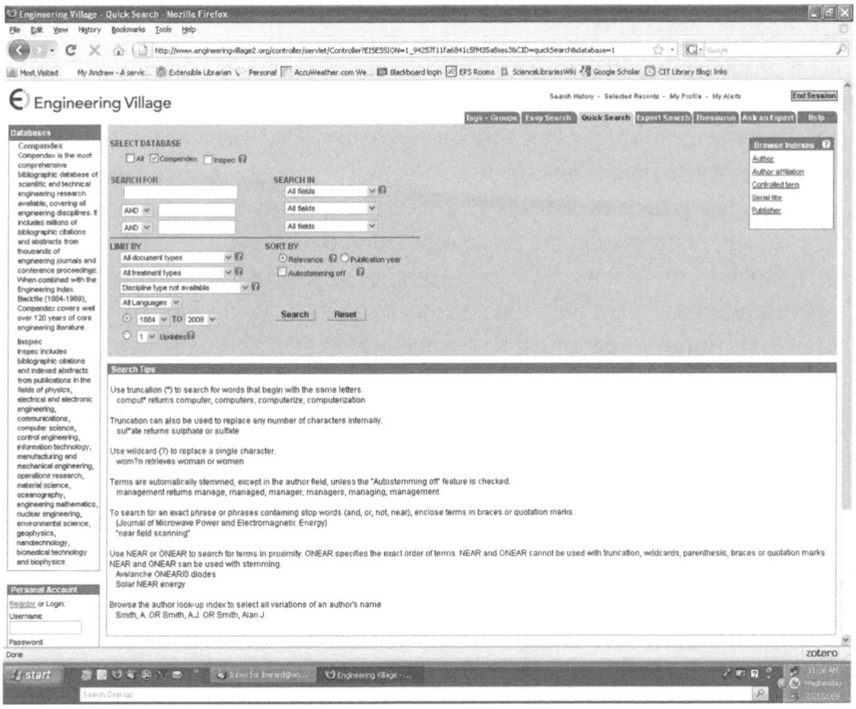

Figure 4.5. El Compendex main search screen. Accessed March 2009.

Figure 4.6. Web of Science main search screen. Accessed March 2009.
Certain data included herein are derived from the Web of Science (r) prepared
by THOMSON REUTERS (r), Inc. (Thomson(r)), Philadelphia, Pennsylvania,
USA: (c) Copyright THOMSON REUTERS (r) 2011__. All rights reserved.

search and these can be printed, e-mailed, or saved to a file for later use. These same results can be saved in the bibliographic style of your choice and uploaded to a citation manager or reference management software on your local computer.

Figure 4.7 represents the results of conducting a "citation search"; this allows a researcher to identify which later articles have cited any particular earlier article, or cited the articles of any particular author, or to determine which articles have been cited most frequently. Citation searching allows a researcher to go both backward and forward in time, where traditional searching only allows for a backward view. As viewed in Figure 4.7, we have input a search using an author's name (Domach, M) and the results page is what we are viewing. From this list it is possible to further select only the variation of the author's name that we wish to view by checking the boxes to the left of each entry. If an entry has the notation "View Record" in the view record column you can click on the highlighted note and examine the full record from the search screen. This "View Record" notation is indicating that this particular citation is linked to an article record within the Web of Science, and an enhanced record will appear for this article that includes links to cited works, additional links to related records, a link to this journal's impact factor rating, and a link to all the references from the article displayed. If

Figure 4.7. Web of Science citation search results screen. Accessed April 2009. Certain data included herein are derived from the Web of Science (r) prepared by THOMSON REUTERS (r), Inc. (Thomson(r)), Philadelphia, Pennsylvania, USA: (c) Copyright THOMSON REUTERS (r) 2011 All rights reserved.

you would like to view the full title of all of the references, click the "Show Expanded Titles" link. It is a general best practice when searching cited references to search the first author's name, as this will provide the most complete results. You need to bear this in mind if there is more than one author for the article, as searching by a secondary or later attributed author may return incomplete results.

Among the valuable tools available on the ISI Web of Knowledge platform is the Journal Citation Reports database. This database lists the journals within a subject field that are considered to have the most influence and research impact within a specialized field. A journal's impact factor is based on quantifiable statistical information based on citation data, in other words, how many researchers cited articles within that journal, thus indicating that it has value within that discipline's global research community. This measure can be of use when making journal collection decisions. As ISI states, "Journal performance metrics offer a systematic, objective means to critically evaluate the world's leading journals" (Institute for Scientific Information [ISI] 2010).

ISI Highly Cited.com, a free product available on the Internet, provides researchers with a gateway to identifying global key players who are making fundamental contributions to the advancement of science and technology within their subject field. Searching across 21 broad subject categories assists one in locating in-depth biographical and publication information for those individuals who have demonstrated great influence in their field. Searches can be conducted by the researcher's name, category, country, or institution.

Scopus, a product of Elsevier Publishing, Inc., has emerged on the scene as the first real competitor to the Web of Science citation database, until recently the only database to offer citation searching (Dess 2006). The most obvious difference between these two citation databases is the years of coverage: Web of Science extends back to 1945 while Scopus begins at 1966.

What Scopus lacks in age it makes up in performance. The presentation of search results is radically different from Web of Science, because in Scopus you search for the author and then view cited references attached to each article rather than viewing a list of all the articles that cite the author in the result set. The speed of search results being returned is impressive and the display and organization of records are clear and easy to interpret. Search results in Scopus contain Web references and patents, a plus over Web of Science. Topic searching is its strength as well as coverage in the scientific, technical, and medical (STM) subject areas. A real weakness here is the lack of coverage of the humanities or the arts and a weak showing in the social sciences unlike the overall disciplinary coverage found in Web of Science. However, from 1995 forward Scopus is a serious competitor to Web of Science in the life sciences and medicine and serves these specialized disciplines very well. At about the same price as Web of Science for most institutions, it can be argued that for comprehensive coverage of STM subject areas, if cost if not a prohibitive factor, the purchase of both databases should be considered.

GENERAL SCI-TECH DATABASES

Applied Science & Technology Abstracts, produced by the H. W. Wilson Company, is just one of the available databases in science and technology for the novice or

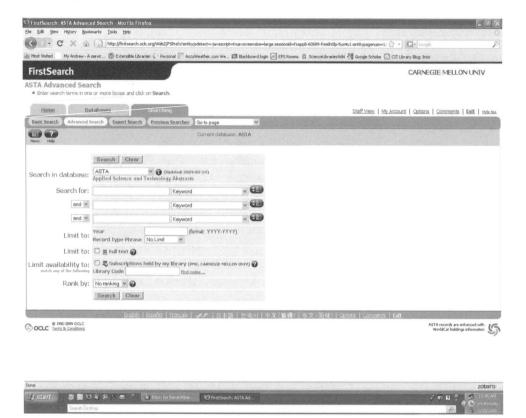

Figure 4.8. Applied Science & Technology Abstracts main search screen from the OCLC FirstSearch platform. FirstSearch is a registered trademark of OCLC Online Computer Library Center, Inc.

layperson. The content is presented in an easy reading format for the nonscientist or technologist. This database contains information in the scientific and technological fields of chemistry, engineering, computer technology, mathematics, data processing, physics, and energy-related disciplines. Searches can be performed by author, keyword, source, and so on and can be restricted to full-text-only results. Many of the articles include career information, trends in science, as well as buyers' guides and directories. This database is an ideal purchase for a high school, public, and general science library. See Figure 4.8 for a view of the main search screen.

REFERENCES

Allison, DeeAnn, Beth McNeil, and Signe Swanson. 2000. "Database Selection: One Size Does Not Fit All." *College & Research Libraries* 61 (January): 56–63.

Armstrong, C. J., Andrew Large, and J. A. Large, eds. 2000. *Manual of Online Search Strategies: Sciences,* 3rd ed. Surry, UK: Gower Publishing Company.

Bell, Susanne S. 2009. *Librarian's Guide to Online Searching: 2nd edition.* Santa Barbara, CA: Libraries Unlimited.

BusinessDictionary.com. 2009. Available at: http://www.businessdictionary.com/definition/born-digital.html.

Cassell, Kay Ann, and Uma Hiremath. 2006. *Reference and Information Services in the 21st Century: An Introduction.* New York: Neal-Schuman Publishers.

Dess, Howard M. 2006. "Database Review and Reports: Scopus." *Issues in Science and Technology Librarianship* (Winter). Available at: http://www.istl.org/06-winter/databases4.html.

Institute for Scientific Information (ISI). 2010. *Journal Citation Reports.* Available at: http://thomsonreuters.com/products_services/science/science_products/a-z/journal_citation_reports.

WEB SITES

Applied Science & Technology Abstracts: http://www.hwwilson.com/Databases/applieds.htm

EI Compendex: http://www.ei.org/databases/compendex.html

Google Scholar: http://scholar.google.com/

ISI Highly Cited.com: http://isihighlycited.com/

The National Library of the Environment: http://www.cnie.org/

Science Citation Index (SCI)/Web of Science: http://thomsonreuters.com/products_services/scientific/ScienceCitationIndex

SCOPUS: http://www.scopus.com/scopus/home.url

Chapter

Current Awareness and Web 2.0 Tools

The standard library *current awareness service* has many definitions. In fact, keeping current on a topic of choice has been an activity engaged in since the 17th century, known by many names, and approached in a variety of ways (Barr 2006). Let's first explore the activity of keeping current and the names and definitions assigned to it.

The purpose of a current awareness service is to inform the users about new acquisitions in their libraries. Among the types of libraries, public libraries are best known for using displays and well-marked shelves as means of drawing attention to new materials. A regular practice in many libraries is the production and marketing of selective lists of newly arrived items. Some libraries have adopted a practice of selective dissemination of information (SDI), whereby librarians conduct regular searches of databases and other resources to find references to new articles or other materials that fit a particular patron's interest profile and forward the results of these searches to the patron (*Encyclopædia Britannica* 2009).

One development of the concept of SDI in the electronic environment is a computer program that scans computer bulletin boards, electronic mail messages, and similar networked information resources and selects items that meet a user profile. Such programs enable individual users to keep abreast of the large amount of information available through computer networks without having to sift through material that may be of little interest or relevance to her or him. This type of automated search service has saved countless hours of work for both the librarian and the researcher.

Current awareness services or SDI are activities that indicate a specialization that has occurred in this core function of libraries and that grew mainly out of the expansion of scientific and industrial research during and after World War II. Three factors strongly influenced this process. First, the increase in research and publication affected all types of libraries and brought with it a similar increase in subject specialization. Second, working scientists, accustomed to referring to reports in published papers, were content to leave the organizing of information searches to a colleague who knew and understood his or her work. Third, the widespread application of scientific research in industry provided an extra stimulus in extracting subject-specific information because of the necessity for speedy application of results to gain commercial success in production. For a historical overview of the birth and developments in current awareness delivery platforms and practice, see Figure 5.1.

In the computer science realm, current awareness is known as a system for notifying users on a periodic basis of the acquisition, by a central file or library, of information (usually literature) that should be of interest to the user (McGraw-Hill Dictionary of Scientific and Technical Terms 2003). Generally the "system" in place would be a computer system; when programmed to search and collect items on a topic, it would then package the results list and send it via electronic messaging.

The provision of current awareness practices and tools can be delivered in a variety of formats. Some examples are:

DEVELOPMENTS IN CURRENT AWARENESS SERVICES

17th Century	18th & 19th Centuries	The 20th Century	The 21st Century....
Creation of the **first scholarly journal** (*Philosophical Transactions of the Royal Society of London, 1660's*) and **professional societies.**	**Formation of subject-specific societies;** led to multiple publications, most notably for a limited subject area and this instituted **the birth of indexing and abstracting journals.**	**Technology developments increase** and the need for searching *forward* as well as *backward* in time because of the unwieldiness of the print format.	**Technology solutions and Web 2.0 Tools** abound and new services spring up to suit the solo researcher's preference.
A tool for scientists and technologists whom wished to be informed on recent developments in their respective fields; provided sustainability of ideas by printing and distributing information broadly.	Many **society libraries were formed** with some created mainly to serve in **the provision of current awareness services.**	**Computers** come to the aid of I & A services; subject databases are created to accommodate the increased rate of publication in sci-tech; SDI services formulated; the **Internet** is invented and user self-service searching becomes the norm.	**Programs include:** RSS (*Really Simple Syndication*) feeds; Table of Contents (*TOC*) sites; Browser searching, (Google, IE, etc.); Blogs; Social bookmarking tools such as Wiki's, Delicious.com, Twitter.com and many more!

Figure 5.1. Developments in current awareness services.
Copyright @ G. Lynn Berard, 2008.

- A database(s) search that contains news and is updated in real time or at least daily
- An SDI topic search against a database within the subject specialty
- A profile that describes the user and what topics she or he wants to receive
- A topic search result delivered in the form of news in full articles or abstracts
- An integrated system that manages the database, topics, profiles, and search results in one coordinated sequence
- E-mail alerts
- Really simple syndication (RSS) feeds
- Blogs

As computerized databases were being established during the 1970s and populated with subject-specific information, the next logical step in knowledge acquisition was the invention of ways to search and mine the data to provide ease of access to the content by category or key term. Below we discuss several services for libraries.

One of the major databases designed specifically for the selected dissemination of information and going back to the 1990s is Current Contents Connect®. This is a current awareness database that provides easy Web access to complete tables of contents, abstracts, bibliographic information, and abstracts from the most recently published issues of leading scholarly journals, as well as from more than 7,000 relevant, evaluated Web sites. Also included is full bibliographic information from some electronic journals before they are published.

Ingenta has traditionally served academic libraries since its inception in 1998, providing free use of the most comprehensive collection of academic and professional publications (http://www.ingentaconnect.com). For a fee, Ingenta allows integration and access of their database to a local library OPAC and can make accessible full-text delivery of documents to the researcher.

OCLC's Electronic Collections Online is a powerful electronic-journal service that offers Web access to a growing collection of more than 5,000 titles in a wide range of subject areas, from over 70 publishers of academic and professional journals. It also provides a robust archiving solution, cross-journal searching, and more.

In today's information climate the use of Web 2.0 tools has taken current awareness capability to new heights and provides an individual with free, easy-to-adapt tools to set up personalized alerts on topics of research interest tailored to his or her specific need. New tools are being designed and implemented at a rapid pace and will have changed by the writing of this chapter.

WEB 2.0 TOOLS

From Wikipedia: The term "Web 2.0" refers to a perceived second generation of Web development and design, that aims to facilitate communication and secure information sharing, interoperability, and collaboration on the World Wide Web. Web 2.0 concepts have led to the development and evolution of Web-based communities, hosted services, and applications such as social-networking sites, video-sharing sites, wikis, blogs, and folksonomies (Cromity 2008).

The term was first used by Dale Dougherty and Craig Cline and shortly after became notable following the O'Reilly Media Web 2.0 conference in 2004. Although the term suggests a new version of the World Wide Web, it does not refer to improvements to any technical specifications, but rather to changes in the ways software developers and end-users utilize the Web.

RSS FEEDS

RSS 2.0 or "really simple syndication" is a term used to describe the standard for the sharing of Web content across different Web sites. It is an XML-based format commonly used for content distribution but it is most often used for the delivery of news headlines on a Web site. Newspapers and other Web content owners who wish to share their information create an RSS document and register that document with an RSS publisher, thereby allowing their content to be read from a site other than their own. Put more simplistically, it acts as a gateway to external sites by gathering links on topics of interest and providing access to those sites all from one Web address. Imagine librarians creating a site to serve clients who are interested in environmental issues; by using RSS feed technology they could gather news feeds, event listings,

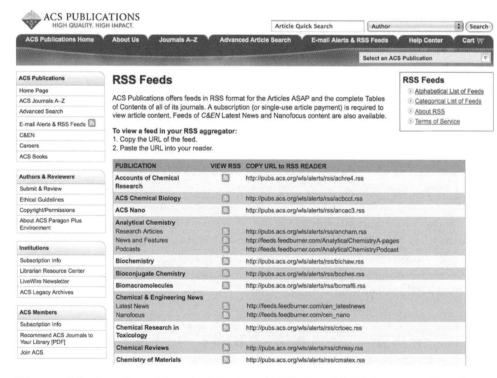

Figure 5.2. American Chemical Society (ACS) list of RSS journal table of contents feeds. Reproduced with permission from the American Chemical Society Publications Division (http://pubs.acs.org/cen/subscribe/rss.html). Accessed 25 February 2010. Copyright 2010 American Chemical Society. All rights reserved.

news stories and major headlines, scholarly articles, and links to institutional information sources and discussion boards, all in one Web site location. This is a 21st-century current awareness tool at work. Let's take a look at an example of RSS technology put to work by a society publisher, the American Chemical Society (ACS).

ACS Publications offers feeds in RSS format for the Articles ASAP service and the complete tables of contents of all of its journals. A subscription (or single-use article payment) is required to view article content. Feeds of *Chemical & Engineering News* (*C&EN), Latest News,* and *Nanofocus* content are also available.

To view a feed in your RSS aggregator:

1. Copy the URL of the feed
2. Paste the URL into your reader

See Figure 5.2 for the American Chemical Society's (ACS) list of RSS journal tables of contents feeds.

Additionally, a technology that can enhance the reading experience for users, called a Web-based reader, allows the selection of news feeds and keeps them all in one place and accessible from any computer. Setting up a personal Web-based reader, such as Google Reader, is a free service open to anyone, and once you have an account, subscribing to news readers available on the Web is fast, easy, and convenient.

BLOGGING

A very popular form of "Web journaling" is the *blog.* A blog (shortened form of "Web log") is a Web site that contains a collection of brief posts, articles, essays, photos, or any other writing or imagery and is authored by an individual or an organization. Blogs can be used for social-networking purposes like staying in touch with family and friends, collaboration as a project site for research or community service groups, networking, or simply as a means of sharing information. Many free blog account programs are available on the Internet, but setting up a blog requires registration and creating a site. Most blogs are open to any reader and invite public comments or posts; however, if privacy is necessary, the site can be password protected and require an administrator to grant permission to individuals as readers and/or authors.

Blog posts and articles are most often brief in nature, chronological, and frequently updated, making them an ideal place for interaction and the sharing of new information and opinions; they also imitate the 19th-century practice of keeping personal diaries and journals. Librarians have embraced this technology as a communication and marketing tool and, along with the creation of career-related blogs, have also implemented blogs in their workplace and in their scholarship endeavors. An example of a scholarly blog is a WordPress blog called the READ Project. This blog was set up as a communication and collaboration tool for participants in a nationwide study in the spring of 2007 in the use of an assessment tool, the Reference Effort Assessment Data or READ Scale. The READ Scale is a six-point scale used for recording supplemental qualitative statistics gathered when reference librarians assist users with their inquiries or research-related activities by placing an emphasis on recording the skills, knowledge, techniques, and tools utilized by the librarian during a reference transaction (see Figure 5.3).

Figure 5.3. A research project management blog for the READ Scale, a reference assessment tool study. Copyright @ Bella Karr Gerlich, 2009.

WIKIS AND GOOGLE DOCS

A *wiki* is a tool for the collaborative creation of a community document, a document that is authored, edited, and modified by the collective endeavors of multiple authors. A Web page in nature, a wiki does not require the user to have any knowledge of specialized authoring languages such as HTML, has very simple formatting, and is a Web-based version of office software. Residing on the Web expands the on-the-fly collaborative capabilities of this software. Presentations, documents, spreadsheets, and forms can be produced on a wiki. Also, wikis are often used in education to enable multiple authoring within online course software programs. See Figure 5.4 for an example of a library-related wiki site.

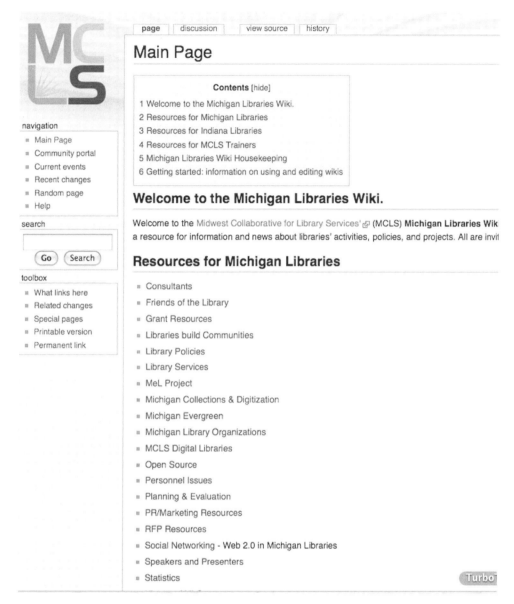

Figure 5.4. A library consortium utilizes wiki technology as a resource and news site for members. Available at: http://mlcnet.org/wiki/index.php/ Main_Page. Copyrighted by Midwest Collaborative for Library Services.

Google Docs, like wikis, enable community authoring but in many respects add on more features for the author and can be used as a free Web-based suite of office software applications.

PUBLISHERS' GATEWAYS (NATURE PUBLISHING)

Many publishers in the science and technology disciplines have embraced the power of the Internet to reach out to their communities and to strengthen their role in enabling scientific research and communication. Nature Publishing (http://www.nature.com) is dedicated to promoting and disseminating advances in scientific endeavor through their Gateways & Databases Web site, a mix of news, analysis, comment, articles, and extensive information tools and resources. The creation of this online community is made possible and enriched by the implementation of a fully Web-based journal publication system, a system that handles and moves scientific papers through the entire editorial process from manuscript submission to peer review and then on to successful publication. Many features have been added in this electronic system such as the delivery of PDFs to the researcher's desktop, the linking of references across all publishers via the digital object identifier (DOI), and the linkage between the original paper and enriched supplementary material added to the online version (Clarke 2007).

With the advent of Web 2.0 and the trend towards enabling interactive discussion capabilities, search tools, and RSS feeds at the publications site, publishers could now offer more innovative ways for research to be quickly disseminated and distilled at one organized Web site. This is the goal of the databases and discussion groups found at Nature's Web site, to provide researchers and novices a one-stop overview of the latest and most significant research in a specialized discipline (see Figure 5.5).

Figure 5.5. Nature Publishing home gateways page (http://www.nature.com). Copyright Nature Publishing, 2009.

Examples of social-networking tools include: Facebook, YouTube, Twitter, Del. icio.us, and LinkedIn. Take a look at the following Web sites to become familiar with just some of the social-networking tools available:

Entry page for Facebook (http://www.facebook.com)
Introduction page for LinkedIn (http://www.linkedin.com)
Entry page for del.icio.us (http://delicious.com)
YouTube home page (http://www.youtube.com)

As we discovered in this chapter, current awareness services have been practiced since the 17th century; yet while our delivery methods and tools have evolved, the basic tenets of service remain the same. Researchers, and the activity of discovery and invention, have always benefited from the efforts of information professionals in locating, selecting, and disseminating found materials in their topic of inquiry. Whether delivered in print or electronic means, SDI activities remain a vital service to the research community.

REFERENCES

Barr, Dorothy. 2006. "Staying Alert: The Wild New World of Current Awareness Services." *C&RL News* (January): 14–17.

Clarke, Maxine. 2007. "Harnessing the Web—Nature's Way: New Tools and Resources to Help Researchers Communicate and Innovate." *Logos* 18 (4): 173–79.

Cromity, Jamal. 2008. "Web 2.0 Tools for Social and Professional Use." *Online* 32 (5): 30–33.

Encyclopædia Britannica. 2009. "Library." Encyclopædia Britannica Online. Available at: http://search.eb.com/eb/article-62059.

McGraw-Hill Dictionary of Scientific and Technical Terms. 2003. s.v. Current Awareness System. Available at: http://www.credoreference.com/entry/mhscience/current_awareness_system.

Chapter

6

Conferences and Society Meetings and Directories

CONFERENCES AND SOCIETY MEETINGS

Each year there are tens of thousands of scientific and technical meetings throughout the world at the local, regional, national, and international levels. These meetings are usually called conferences, congresses, or symposia and range from small, specialized gatherings to very large international conventions. The term "symposium" (singular) comes from the Greek roots meaning "drinking together." In ancient Greece, a symposium was, literally, a drinking party at which there was intellectual conversation. The majority of scientific and technical societies organize annual meetings at which the results of original research are presented. Indeed, one investigation reported that researchers reported the results of laboratory investigations more frequently at national meetings than at any other type of meeting they attend (Garvey, Lin, and Nelson 1970). In another study that examined the role of the national meeting in the scientific communication process, it was determined that the national meeting has developed a distinct and important function in this process. Specifically, the national meeting is the earliest occasion for the early release of the results of laboratory work and the last major informal medium before such work is published (Garvey et al. 1972). Today, a symposium, or more frequently the conference, is a formal meeting or gathering to discuss some particular subject where the results of research are presented and ideas are freely exchanged.

Conferences provide a variety of functions and activities including:

* Announcement of new information including the newest and latest equipment
* Formal and informal exchange of information
* Identification of interdisciplinary topics and newly emerging fields for investigation

- Professional growth and development activities, including continuing education
- Association activities including fact finding, reporting, policy formulation, and governance

As a subject, conferences and their growth over time have been the focus of several investigations. In one study, it was reported that there were 3 international conferences in 1853, more than 100 in 1909, and approximately 2,000 in 1953 (King 1961). In that same decade of the 1950s, another investigation estimated the number of international conferences per year to be 5,000 (Murra 1958). A more recent study provides a figure of 10,000 conferences held annually (Short 1972). Whatever the actual number is, the number of conferences organized each year appears to be increasing due to such factors as: (1) increase in the amount of laboratory research, (2) increase in scientific and technical information as a direct result of this research, (3) willingness to share this new information, and (4) the need to communicate this information more quickly than by traditional routes, particularly journal publication.

In addition to tracking the number of conferences held over time, attendance at conferences has also been studied. In one detailed investigation, attendance at the annual meetings of the Federation of American Societies for Experimental Biology (FASEB) grew from approximately 2,000 participants in 1942 to over 15,000 in 1963 (Orr, Coyl, and Leeds 1964).

The significance of presenting and disseminating scientific information at conferences has been recognized as an important means of communication in science and technology, especially since World War II. Papers presented at conferences and scientific meetings are important for current awareness because of the currency of the information reported in them. They are also a valuable source and survey of emerging and developing subjects including interdisciplinary topics.

In addition to the publication of conference papers in the traditional print format, the Internet has enabled conferences of all sizes and in numerous subject areas to create their own Web sites and gather valuable information, including full-text and Power-Point presentations in one place. In today's rapidly changing technological environment, there is quick and easy access to conference papers. The open access movement has permitted free and immediate availability of some types of conference literature.

Through institutional repositories, conference papers are almost immediately available to the public. Personal contact information for the conference presenters can be located at their respective institutions or on social-networking Web sites, allowing for one-on-one communication and information exchange.

Conferences have a variety and wide range of sponsors. Many are sponsored by learned and professional societies such as the American Chemical Society, the American Institute for Biological Sciences, or the American Society of Mechanical Engineers and involve national and regional groups. Other conferences may be international and sponsored by such groups as the United Nations or the International Council of Scientific Unions. Government agencies, particularly at the national level, sponsor conferences, often in response to emerging or ongoing issues and problems. Oftentimes, there is no society or governmental sponsor for a conference and sometimes entire industries, private companies, and foundations sponsor and support conferences and related meetings.

Information on current and future meetings and conferences should provide the following elements at a minimum:

- Sponsoring organization
- Name or topic of the conference
- Inclusive dates
- Location
- Name of the contact person for additional information, including registration

The conference literature consists of the papers presented at meetings and conferences and as a primary source of information is very valuable to the scientist or engineer. In most instances, information presented at conferences is the first public disclosure of the results of laboratory work. The information resulting from original research is usually presented at conferences and similar meetings many months before its appearance in more formal outlets including the periodical literature. In addition, the comments and discussion generated by such presentations play important roles in both the research process and scientific communication. At this point in the research endeavor, communication and discussion may elicit feedback that could ultimately lead to modifications in experimental design, laboratory procedures, or data interpretation.

Conferences are particularly amenable for interaction and discussion among researchers as well as exhibitors. Although the presentation of formal papers is the major activity of conferences and symposia, much of what is important to an individual researcher occurs outside these presentations. New information, new ideas, and new insights directly related to an individual's current laboratory research obtained from colleagues are frequently more valuable that the information within the contributed papers. In rapidly changing fields, such as computer science, much of what is known is only available in print and electronic conference literature. Only a small percentage of the information in these rapidly advancing fields appears in the journal literature of these fields. Furthermore, papers presented at conferences may not have been subjected to the same requirements and standards in the review process that is generally required for acceptance and publication of a journal article.

Conference planning and implementation require a significant amount of work. Long before the actual conference, the real work begins. Announcements, calls for papers, preliminary programs, and other forms of advertising are published in subject-specific journals as well as general ones such as *Science* and *Nature*. Abstracts of papers (including programs and other relevant information) to be presented at a conference are often distributed or downloaded before the actual conference to those individuals who have registered to attend. In other instances, the abstracts and other information such as key papers may be published in print or electronic format in journals, newsletters, or similar outlets prior to the actual conference. This type of information is generally thought of as the "preconference literature" and frequently results in the generation of a significant amount of print or electronic information.

Other types of literature are also available during the actual conference. A detailed program and additional information including abstracts or preprints of selected or all papers are usually available to all who have pre-registered or those who register on-site. In addition, other information such as the opening and closing addresses, keynote speeches and presentations, business resolutions, and related materials are frequently available in a timely manner in a variety of formats during the actual conference.

The "postconference" literature may appear in one of several forms in either print or electronic format. The major forms of publication for the proceedings of conferences and society meetings include:

- A single or multivolume work containing the entire proceedings of a conference or meeting
- A monograph or report with a specific title and editor, including as a volume with an ongoing monographic series
- A supplement, special number, or entire issue of an established journal because it is one of the official publications of the society or agency that organized the conference, or because the subject content of the conference was deemed important to the society or agency that published it
- Selected papers or abstracts published in a journal because it is the official organ of a society or agency or because of the importance of the subject content
- Reports of a conference or a meeting in a journal that has a special section devoted to "congress or conference proceedings"
- Simultaneous publication as an issue or part of a journal and as a monograph or report (Cruzat 1967)

As a result of this wide variation in publication format, it is frequently difficult to identify, verify, and locate material (i.e., papers) presented at conferences.

It has been estimated that more than 150,000 papers related to engineering are published each year (Moline 1988). Many of these papers will have initially been presented at a scientific meeting or conference, whether it is a routine report of research or a breakthrough of major importance. However, a significant number of papers read at conferences are never published and are permanently lost to the body of scientific literature and the scientific community. According to one estimate, as much as 25 percent of conference and symposium papers may never appear in print (Short 1972). In another study, the abstracts presented at conferences were tracked for subsequent publication. Of almost 400 presentations, slightly more than 50 percent (51.5%) were published while 48.5 percent were not published, and of the abstracts that were published, about one-third (33%) appeared in journals other than that in which the abstract had originally appeared (Liebesny 1959). Another study investigated the timeliness in the publication of the overall (i.e., the complete) conference proceedings. Of 194 conference proceedings that were studied, 42 were published in the same year that the conference was held, 93 were published in the following year, 46 were published two years later, and the remaining 13 appeared three or more years after the conference was held (Hanson and Janes 1960).

In the current online environment, some aspects of the conference literature have improved, such as the speedy and easy access to conference papers including supporting materials such as PowerPoint presentations. However, there are ongoing problems including delays in publishing the final conference papers, lack of coverage of conferences and their content in major databases, difficulty in locating publishers of conference materials, and the failure to publish some of the conference literature. It has been estimated that as much as 90 percent of the material that ultimately ends up as a published journal article has been previously disseminated to individuals within a somewhat narrow research area (Garvey 1979). It is generally acknowledged that in some areas such as physics and computer science, the preprint (print at first, but now almost exclusively electronic) continues to serve as the preferred informal method to communicate with a network of peers.

Although the literature generated as a result of conferences, symposia, and society meetings is very important, there are a number of problems associated with this type of literature. The publication of the final proceedings may be delayed due to the time involved in obtaining manuscripts from speakers, editorial processing, and the actual production time of the final volumes whether in print or electronic format. Conference volumes may contain material that has already been published, such as that already provided to registrants. In addition, conference volumes may contain contributions that would not have been approved by referees for publication in a journal and this material may not contain adequate bibliographic references. Finally, it is frequently difficult to identify, locate, and retrieve information that is embedded in conference and symposium volumes.

Information on future meetings and conferences can be obtained from a variety of sources. The typical three sources are:

- Scientific Meetings. San Diego, CA: Scientific Meetings Publications, 1957–.
- Mind: The Meetings Index. InterDok Corporation, 1983– (www.interdok.com/mind).

  - This site offers free online access to locate future conferences, congresses, meetings, and symposia in four subject areas: (1) science/technology, (2) medical/life sciences, (3) pollution control/ecology, and (4) social sciences/humanities.

- International Congress Calendar. Brussels, Belgium: Union of International Associations, 1960–. Available electronically since 2004 from UIA (www.uia.be/calendar).

  - This publication is a comprehensive source of information on past and future international meetings organized or sponsored by international organizations, both governmental and nongovernmental. The database contains almost 330,000 events dating from 1986 and more than 15,000 future events are included each year. The content includes meeting name, date range, theme, location, organizer names, contact information, and participant numbers and associated exhibition when available. Current and past congresses are searchable geographically, chronologically, and by keywords of the event theme.

The following sources are examples of those that are frequently useful in locating conference papers and proceedings:

- Conference Papers Index. Bethesda, MD: CSA, 1982–. Available through ProQuest.

- This publication provides citations to papers and poster sessions presented at major scientific meetings worldwide. Subject emphasis since 1995 has been in the life sciences and environmental sciences, while older material also covered the physical sciences and engineering. Information is gathered from final programs, abstracts, and published proceedings as well as from questionnaire responses. Records include complete ordering information to obtain preprints, abstracts, proceedings, and related publications derived from the conference, together with author and title information needed to track specific papers from the conference.

- Conference Proceedings Citation Index.

  - This title is accessed via Web of Science and extracts 30 percent of its data from journals and 70 percent from books and covers the most significant conferences, symposia, seminars, colloquia, workshops, and conventions worldwide.

- Conference Proceedings Monthly (CPM). Red Hook, NY: Curran Associates, 2009– (www.proceedings.com/emediatoprint.html).

  - In late 2008, Curran Associates completed their purchase of InterDok Corporation's Acquisition Services business, which had published the Directory of Published Proceedings (DoPP) since 1965. CPM is an electronic listing of thousands of new conference titles per year, including over 1,000 recurring conferences. Various societies have agreed to provide Curran the rights to print from CD or other electronic format for those individuals and institutions that still prefer print.

- PapersFirst. OCLC, 1993–.

  - OCLC is a non-profit, membership-driven, computer library service and research organization dedicated to the public purposes of furthering access to the world's information and reducing information costs. This is an OCLC index of papers presented at conferences worldwide. It covers and provides citations of every published congress, symposium, conference, exposition, workshop, and meeting received by The British Library Document Supply Centre. Contained in each record is a list of the papers presented at each conference.

- ProceedingsFirst. OCLC, 1993–.

  - This index of worldwide conference proceedings is produced by OCLC. It covers and provides citations of every published congress, symposium, conference, exposition, workshop, and meeting received by The British Library Document Supply Centre. Contained in each record is a list of the papers presented at each conference. There are over 169,000 records in this database.

- Web of Conferences (www.webofconferences.org/).

  - EDP Sciences is a publishing house owned by learned societies and run by publishing professionals. It receives support from several scientific communities and provides technical and sales support for the launching and growth of scientific or technical journals or professional magazines on an international scale. EDP Sciences launched Web of Conferences, a platform entirely devoted to scientific conferences. It provides to scientists and conference organizers: (1) an international meeting calendar to announce or find an event, (2) a space that collects all the proceedings published by EDP Sciences, and (3) editorial services for the publication and distribution of conference proceedings that are suitable to any conference, whatever the number of participants and the type of documents supplied. Publishing options for the conference proceedings include online publication, open access, print, and CD/DVD format.

In the past decade there have been an increasing number of conferences that are available in the virtual world. The number and complexity of these meetings continue to grow as organizations and institutions are advertising and presenting such programs. These virtual meetings began in the business world as a result of cutbacks in travel expenses, while at the same time advances in communication technology were exploding. With less time and less travel money available to the research scientist, as well as more and more conference choices, together with better technology, conference and meeting planners are faced with challenges including physical as well as virtual attendance. However, non-virtual conferences are still one of the few environments where researchers can enjoy face-to-face human contact and gather to discuss shared interests and topics at length as well as network with colleagues. As scientists have less and less time for activities besides actual laboratory research, and as more and more conferences are held, planners are struggling with the challenges facing the industry, especially as the technology associated with these meetings and conferences continues to evolve.

Whatever type of conference a scientist chooses to participate in or attend, an important consideration is the value of the conference itself. Interestingly, a recent study funded by the Bill and Melinda Gates Foundation provided some quantitative assessment of the impact of attendance at conferences (Aiken 2006). Specifically, the issue of time and money saved back at the laboratory based on information gathered at the conference was assessed by two groups of attendees. The average amount of time saved in the lab ranged from 2 to 33 weeks and the average amount of money saved ranged from $1,000 to $48,000. The medians were 6 weeks time saved and $6,000 saved. Attendees should and can expect such value from attendance at a conference, but there are other benefits, including a young scientist learning from an experienced leader in the field as well as any attendee making contacts that lead to the efficient sharing of resources. Attendance in either form will, hopefully, lead to accelerated discoveries that can ultimately benefit society.

DIRECTORIES

Many conferences and meetings are sponsored by associations and related societies. Directories are an important type of reference publication and membership directories are frequently published by associations. More and more associations are compiling and maintaining online directories of members that are easy to update and can be searched by association members. The three types of directories (biographical, organization or company, and product directories) are basically lists usually arranged in alphabetical order for easy use and generally provide a limited amount of information. There may be additional finding aids such as a separate geographical index or one that offers membership subject expertise, including members who are consultants in one or more areas. Information in directories changes frequently and they are often updated by publishing supplements or entirely new editions.

Biographical directories are useful for locating information on individuals. However, they are becoming more and more difficult to locate in the printed format. One of the few titles that is still available in print format is *Directory of Physics, Astronomy and Geophysics Staff* published by the American Institute of Physics. This source provides information on more than 36,000 scientific staff in approximately 2,600 institutions and organizations within the physics, astronomy, and geophysics community. The information is divided into institutional and individual listings. Each institutional listing provides the address, phone, fax, e-mail, URL, and current faculty or staff. Individual listings detail each person's institutional affiliation, along with their work address, phone, and e-mail. The directory contains academic physics, astronomy, geophysics and related departments, as well as two-year colleges. It also includes private firms, industrial research and development (R&D) centers, small R&D companies, federally funded R&D centers, government agencies, university-affiliated and other research institutes, and professional societies serving the physics and physics-related communities around the world. Another example of an important and ongoing biographical directory, available in both print and electronic formats, is the *American Chemical Society Directory of Graduate Research*. It is a comprehensive source of information on chemical research and investigators at colleges and universities in the United States, Canada, and Mexico. It covers close to 700 academic departments, nearly 10,000 faculty members, and cites almost 65,000 publications of these researchers. The directory includes listings for chemistry, chemical engineering, biochemistry, medicinal and pharmaceutical chemistry, polymers and materials science, and related areas. In the Web version, the search for faculty includes specific research area, academic rank, gender, and state, with the results producing complete faculty contact information including direct links to faculty e-mail addresses and Web pages, when provided. Institutional searches provide all departmental contact information together with statistical data on the number of faculty and students in the department and a complete list of faculty active in graduate research.

Organization or company directories provide information on academic and research institutions, government agencies, or private sector companies. The type and amount of information varies widely, ranging from a minimum of name, address, telephone number, fax number, and e-mail address to detailed descriptions of research and sales activities, products, personnel, and financial information. Examples of organization or company directories include:

- *Encyclopedia of Associations.* 48th ed. Kristy A. Harper, Project Editor. Detroit, MI: Gale Cengage, 2009. ISBN: 9781414440286.

  - This is the primary source of information on more than 25,000 American organizations of national scope. Entries include the organization's complete name, address and phone number, primary official's name and title, fax number, founding date, purpose, activities, dues, publications, and conferences. A companion set, *Encyclopedia of Associations: International Organizations,* provides information on almost 32,000 international associations from 210 countries. Both titles are available in print and electronic formats.

- *Research Centers Directory.* 37th ed. Donna Wood, Project Editor. Detroit, MI: Gale Cengage, 2009. ISBN: 9781414421599.

  - This directory provides information on more than 15,000 university-related and other non-profit research organizations in the United States and Canada. Research includes fundamental, applied, and developmental studies, as well as data collection, analysis, and synthesis. It is a comprehensive guide to the programs, staffing, publications, and services of research centers such as laboratories, institutes, experiment stations, farms, think tanks, technology transfer centers, research parks, and similar research facilities. The *International Research Centers Directory* covers more than 13,000 research units in 160 countries. Both titles are available in print and electronic formats.

- *Government Research Directory.* 25th ed. Donna Batten, Project Editor. Detroit, MI: Gale Cengage, 2009. ISBN: 9781414434780.

  - This comprehensive source provides information on more than 7,500 research programs and facilities associated with the U.S. and Canadian federal governments. Programs in basic, applied, exploratory, developmental, theoretical, and experimental research are included in all disciplines from astronomy to zoology. Entries provide the usual information and also include Web URL and homepage.

Product directories list products, particularly manufactured goods. They frequently contain information on manufacturing and distributing agencies together with personnel information. One example of a product directory is:

- *Grey House Safety & Security Directory.* Amenia, NY: Grey House Publishing, 2009. ISBN: 9781592373758.

  - Published continuously since 1943 as *Best's Safety Directory,* this title was acquired in 2002 by Grey House Publishing. It is a comprehensive product directory and buyer's guide, which includes almost 7,000 safety products and service descriptions for the safety and security industry. Arranged by topic (e.g., electrical lighting and safety, fall protection, nose and vibration, protective equipment and apparel, etc.), each chapter contains Occupational Safety and Health Administration (OSHA) regulations, training articles, self-inspection checklists and safety guidelines for that topic, as well as product descriptions and product listings in the buyer's guide. It is thoroughly indexed with six indexes ranging from a geographical index of manufacturers and distributors to a brand name index. This directory is available in print and online.

Other titles in this category include *Lockwood-Post Directory of Pulp & Paper Mills, World Aviation Directory,* and *Thomas Register Directory,* which is frequently referred to as the most comprehensive resource for finding information on suppliers of industrial products and services in North America.

Directories are important and easy to use reference tools that list information for people, organizations, and products. There are thousands of directories in all areas of science and technology. Print directories are generally expensive to produce and maintain and are quickly dated. As a result, many print directories do not survive because they are too costly for purchase on a regular basis. Online directories are an alternative to the print product that allow for more frequent updates in a timely manner while providing more recent data and saving time and money.

REFERENCES

Aiken, James W. 2006. "What's the Value of Conferences?" *The Scientist* 20 (5): 54–56.

Cruzat, Gwendolyn S. 1967. "Keeping Up With Biomedical Meetings." *RQ* 7 (Fall): 12–20.

Garvey, William D. 1979. *Communication: The Essence of Science: Facilitating Information Exchange among Librarians, Scientists, Engineers, and Students.* New York: Pergamon.

Garvey, William D., Nan Lin, and Carnot E. Nelson. 1970. "Communication in the Physical and the Social Sciences." *Science* 170 (3963): 1166–73.

Garvey, William D., Nan Lin, Carnot E. Nelson, and Kazuo Tomita. 1972. "Research Studies in Patterns of Scientific Communication: II. The Role of the National Meeting in Scientific and Technical Communication." *Information Storage and Retrieval* 8 (4): 159–69.

Hanson, C. W., and M. Janes. 1960. "Lack of Indexes in Reports of Conferences: Report of an Investigation." *Journal of Documentation* 16 (2): 65–70.

King, Alexander. "Concerning Conferences." 1961. *Journal of Documentation* 17 (2): 69–76.

Liebesny, Felix. 1959. "Lost Information: Unpublished Conference Papers." In *Proceedings of the International Conference on Scientific Information, Washington, D.C., Nov. 16–21, 1958,* 475–79. Washington, DC: National Academy of Sciences–National Research Council.

Moline, Gloria. 1988. "Secondary Publisher Coverage of Engineering Conference Papers: Viewpoint of Engineering Information, Inc." *Science & Technology Libraries* 9 (2): 47–61.

Murra, Kathrine O. 1958. "Futures in International Meetings." *College and Research Libraries* 19 (6): 445–50.

Orr, R. H., E. B. Coyl, and A. A. Leeds. 1964. "Trends in Oral Communication among Biomedical Scientists: Meetings and Travel." *Federation Proceedings* 23 (5P1): 1146–54.

Short, P. J. 1972. "Bibliographic Tools for Tracing Conference Proceedings." *IATUL Proceedings* 6 (2): 50–53.

Chapter

7

Dictionaries and Encyclopedias

DICTIONARIES

Dictionaries are essentially word books and are among the most frequently used reference titles. They are usually arranged in simple alphabetical order. At a minimum, dictionaries provide the definitions of words, phrases, abbreviations, and other terms. In addition, they may provide etymology, pronunciation, usage, field of science or engineering that the word is associated with, or illustrations. Some dictionaries may contain historical information, including names, birth and death dates, and important contributions of an individual to a specific field of science.

Although many science and technology dictionaries rarely give more than the meaning of terms, they may vary significantly in how thoroughly the terms are defined. Entries may consist of one line or be as extensive as several paragraphs. It is not unusual for definitions to be lengthy and detailed in some dictionaries.

In the past, dictionaries were only available in printed format. Today, dictionaries are available in a variety of formats including:

- Printed dictionaries
- Online dictionaries
- Dictionary programs

This chapter offers textual examples of the above categories together with visual examples. Print format has been around since the invention of the printing press, with the oldest printed dictionary compiled as a Latin-English "wordbook" by Sir Thomas Elyot and published in 1538. The first commercially printed dictionary in the United States appeared circa 1806 and was authored by Noah Webster. Webster published *A Compendious Dictionary of the English Language,* the first truly American dictionary, and went on to author an expanded edition, *An American Dictionary of the English*

Language, for which he learned 26 languages in order to research the origins of his own country's tongue. This edition, published in 1828, was a monumental work and contained 70,000 entries. Webster's new work was touted as having exceeded Samuel Johnson's 1755 British masterpiece, *Dictionary of the English Language,* not only in scope but in authority as well.

Online dictionaries can be subscribed to as single-purpose reference resources or can be one of many reference resources accessible via a multidisciplinary/multipurpose reference resource database. One such service is called Credo Reference and is produced by a vendor of the same name. Credo has partnered with numerous publishers to bring together in one product over 380 reference works comprising over 3.1 million entries. A service of this kind is available by contract and a fee is set for annual or multi-year time periods. Depending on the type of library subscribing, the fee may be set by enrolled students (FTE) for academic settings, or by other community measures for public and private institutions.

Many single-purpose generalist dictionaries are available for free on the Internet and often include more than a word definition and may offer thesaurus and phonic elements. It is less frequent to find high-quality free scientific and technical dictionaries online but not impossible. An example of a free online Web site, funded by sponsors, is LibrarySpot.com. This free virtual library resource center has been created by a team of editors working out of Northwestern University/Evanston Research Park in Evanston, Illinois, and its goal is "to break through the information overload of the Web and bring the best library and reference sites together with insightful editorial in one user-friendly spot" (LibrarySpot.com 2009, accessed in May).

Dictionary programs are user-friendly programs that can help with writing tasks and generally are purchased for individual use and loaded on personal computers. Using a program like this you can search for words by standard definition, synonym, homophones or words that rhyme, as well as search out words that can be used to form crosswords. Other features generally available include verbal illustrations, pictures, audio pronunciations, as well as the origin of root words. Dictionary software programs can cost between $20 and $120 or more. As an example, Babylon Pro is one of the more costly dictionary platforms but contains access to a number of dictionaries and thesauri from which to choose as well as features such as instant updates of new material on the fly.

A bilingual or translation dictionary is a specialized dictionary used to translate words or phrases from one language to another. Bilingual dictionaries can be *unidirectional,* meaning that they list the meanings of words of one language in another, or they can be *bidirectional,* allowing translation to and from both languages. Bidirectional bilingual dictionaries usually consist of two sections, each listing words and phrases of one language alphabetically along with their translation. In addition to the translation, a bilingual dictionary usually indicates the part of speech, gender, verb type, and other grammatical information to help a non-native speaker use the words. Other features sometimes present in bilingual dictionaries are lists of phrases, usage and style guides, verb tables, maps, and grammar references. Figure 7.1 shows a sample page from *Patterson's German-English Dictionary for Chemists.*

In the sciences, a large number of dictionaries are multilingual. In such a dictionary, the most common format is to have the English-language term listed, followed by all the foreign equivalents on the same line or in a column. Usually the definition is color-tüchtig *a* capable of having color

Columba-säure *f* columbic acid (from calumba)

Columb-eisen *n* (*Min.*) columbite

Columbia-kopalinsäure *f* columbiacopalinic acid. -kopalsäure *f* columbiacopalic acid

Columbo-wurzel *f* calumba (root)

Colza-öl *n* colza oil

Compakt-platte *f* compact disc

computer-gesteuert *a* computer-controlled. computerisieren *vt/i* computerize. computer-unterstützt *a* computer-assisted

conaxial *a* coaxial

conchieren *vt* shell (out)

Condurit *n* conduritol

conglobieren *vt* heap up

Congo-farbe *f* Congo color (*or* dye)

Conidien- (*Bot.*) conidial

Conifern-harz *n* fir resin

Coniin *n* conine, coniine

Conima-harz *n* conima (resin)

conphas *a* in phase

Conspersa-säure *f* conspersic acid

Constitual-kampf *m* environmental struggle

Continü-küpe *f* (*Dye.*) continuous vat

Copaiva-balsam *m* copaiba (balsam). -öl *n* oil of copaiba. -säure *f* copaivic acid

Copoly-addukt *n* addition copolymer. -harnstoff *m* urea copolymer. -kondensat *n* copolymer. -kondensation *f* condensation copolymerization

copolymer *a* copolymeric. Copolymerisat *n* (-e) copolymerized product

Cops-färberei *f*, -färbung *f* cop dyeing

Corduan-leder *n* cordovan (leather)

Cornicular-säure *f* cornicularic acid

Corozo-nuß *f* ivory nut

corr. *abbv* (corrigiert) corrected; proofread

Cörulignon *n* cerulignone

Cosekante *f* (-n) cosecant. Cosinus *m* (..nen) cosine

Costus-säure *f* costic (*or* costusic) acid

Cotangente *f* (-n) cotangent

Cotarn-säure *f* cotarnic acid

Coto-rinde *f* coto bark

Cotta *f* (*pl* Cotten) (*Metll.*) bloom, lump, coke

cottonisieren *vt* cottonize

Cotton-öl *n* cottonseed oil

Couch-tisch *m* coffee table

Couleur *f* (-en) 1 color. 2 caramel; burnt sugar; dark coarse smalt

Coulomb-feld *n* coulombic field

coupieren *vt* cut (short), etc = KUPIEREN

Coupüre *f* (-n) (*Tex.*) reduced print; reduction of print paste

couragiert *a* courageous

Covellin *n* (*Min.*) covellite

C-Polymerisation *f* condensation polymerization

crabb-echt *a* (*Tex.*) crabbing-resistant

Crack-benzin *n* cracked gasoline

cracken *vt* (*Petrol.*) crack. Cracken *n* I cracking. II crackene. Cracken-chinon *n* crackenequinone

Crack-kessel *m* cracking retort. -prozeß *m* cracking process

Crackung *f* (-en) cracking (of petroleum). Crack-verfahren *n* cracking process

Craig-verteilung *f* countercurrent distribution

creme *a* cream(-colored) = KREM. Creme *f* (-s) cream = KREM

Crêpe-kautschuk *m/n* crepe rubber

Cresyl-säure *f* cresylic acid, cresol

Cribia-teil *m* phloem element

Crocein-säure *f* croceic acid

Croton-alkohol *m* crotyl alcohol. -harz *n* croton resin. -säure *f* crotonic acid

Croupon *m* (-s) (*Lthr.*) crop, butt. crouponieren *vt* crop, round (hides)

C-Stoff *m* rocket fuel

Cubeben I *n* cubebene. II cubebs: *pl of* Cubebe *f*

Cubeben-öl *n* cubeb oil. -pfeffer *m* cubebs. -säure *f* cubebic acid

Cuite-seide *f* boiled-off silk

Cumalin *n* coumalin. --säure *f* coumalic acid

Cumar-aldehyd *m* coumaraldehyde (o-hydroxycinnamaldehyde)

Cumaril-säure *f* coumarilic acid

Cumarin-säure *f* coumarinic acid

Cumaron-harz *n* coumarone resin

Cumar-säure *f* coumaric acid

Cumin-öl *n* cumin (seed) oil. -samen *m* cumin seed. -säure *f* cum(in)ic acid

Figure 7.1. Sample page from *Patterson's German-English Dictionary for Chemists*, 4th ed. By Austin M. Patterson; edited by George E. Condoyannis. New York: Wiley, 1992. Reprinted with permission of John Wiley & Sons, Inc.

omitted and such dictionaries are often referred to as "word equivalent dictionaries." Elsevier Publishing Company has published many multilingual dictionaries. In *Elsevier's Dictionary of Chemical Engineering,* the six equivalent languages include: English, French, Spanish, Italian, Dutch, and German. Other subject dictionaries may include Japanese or Russian as a substitute for one of the languages. Figure 7.2 shows a sample entry from a multilingual dictionary in chemical engineering. Other multilingual titles provide more detail than word equivalents. In *Elsevier's Dictionary of General Physics,* each entry is numbered and includes the English word or phrase together with the subject area (e.g., crystallography) and definition. Following the definition is the equivalent word or phrase from the foreign languages included. In the general physics dictionary, the languages included in addition to English are French, Spanish, Italian, Dutch, and German. Figure 7.3 shows a sample entry from a multilingual dictionary that includes definition and subject area for a particular term.

Thesauri are often confused with dictionaries. They are different from dictionaries in that they do not contain definitions or explanations of words. Thesauri are controlled vocabularies that display relationships among terms in a scientific or technical discipline to facilitate indexing and retrieval of documents. Typically a thesaurus focuses on one discipline or field of study, such as metallurgy or the environment.

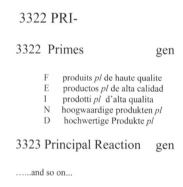

3322 PRI-

3322 Primes   gen

    F   produits *pl* de haute qualite
    E   productos *pl* de alta calidad
    I   prodotti *pl* d'alta qualita
    N   hoogwaardige produkten *pl*
    D   hochwertige Produkte *pl*

3323 Principal Reaction   gen

……and so on...

Figure 7.2. A sample entry from a multilingual dictionary in chemical engineering.

Aerosol

No.	English GB *and* US	Subject	Definition	Francais	Italiano	Deutsch
100	**aersol**	phc	A colloidal system in which a gas, usually air, is the continuous medium, and particles of solid or liquid are dispersed in it.	aerosol *m*	aerosole *m*	Aerosol *n*
101	**aerostatics**	phc	The science of gases at rest (mechanical equilibrium).	aerostatique *f*	aerostatica *f*	Aerostatik *f*
102	**after-flow**	mec	Persistence of flow after removal of the external stresses, due to relaxation of visco-elastic stresses.	persistence *f* De l'ecoulement	persistenza *f* del flusso	Nachstrom *m*

Figure 7.3. A sample entry from a multilingual dictionary that includes definition and subject area for a particular term.

Terms are generally not defined except for brief scope notes and synonyms. Thesauri most frequently display hierarchical (i.e., generic to specific) relationships among scientific and technical terms. These hierarchical relationships are used to indicate terms that are narrower or broader in scope. In metallurgy, "metal" would be a broader term (BT), whereas "copper" would be a narrower term (NT). Other relationships are use (USE), used for (UF), and related term (RT).

Electronic versions of thesauri provide the same forms of hierarchical relationships of terms as their printed parents traditionally always have, but many now offer other services as well, all on the same online screen. Online thesauri often expand their usage by embedding dictionary, phonic, and article-searching capability all in one search form, thereby reducing the types of reference materials required on the researcher's desktop and reducing the need for multiple transactions. Online database services, such as Engineering Information, now include thesaurus tools in their products, adding support for building an accurate search strategy. Figure 7.4 shows a screen shot from the *Engineering Index* online thesaurus.

While there are numerous dictionaries in science and technology, three titles that are representative of the general science and technology dictionaries include:

1. *Academic Press Dictionary of Science and Technology.* Edited by Christopher Morris. San Diego, CA: Academic Press, 1992. ISBN: 0122004000.

   With more than 2,400 pages of information, this title has more than 130,000 fully defined entries (not counting abbreviations) and, while somewhat dated, is still one of the best science and technology dictionaries that has ever been published. First published in 1992, it was issued

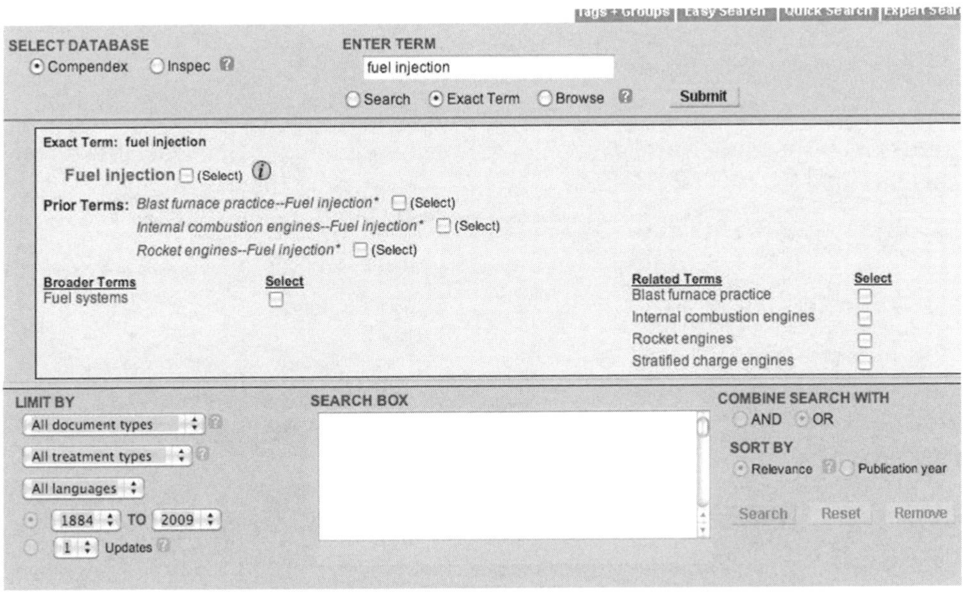

Figure 7.4. A screen shot from the *Engineering Index* online thesaurus. Copyright © 2009 Elsevier Inc.

in 1996 as a CD-ROM product to provide additional search capabilities in digital format. At the time of publication, it was considered the new standard of excellence in science and technology dictionaries. It covers 124 fields of science and engineering from acoustics to zoology and includes pronunciation guides for difficult or phonetically irregular terms. Each entry has 1 of the 124 scientific fields assigned to it to further help in understanding the definition. Cross-references link overlapping and new fields and more than 2,000 illustrations and photographs, including a dozen color illustration sections. An extensive appendix includes frequently consulted scientific data (symbols and units, physical constants, atomic weights, etc.), the periodic table of the elements, standard weights and measures, and a chronology of science. One unique feature of this title is the "windows definition." For each of the 124 disciplines covered, a recognized leader in the field has written a brief overview and synopsis of the area. It is now available online through Credo Reference (http://www.credoreference.com/), first mentioned in our introduction to this chapter. Figure 7.5 shows a screen shot from *Academic Press Dictionary of Science and Technology,* available on the Credo Reference platform.

2.   *Chambers Dictionary of Science and Technology.* General Editor, Peter M. B. Walker. London: Chambers, 1999. ISBN: 0550141103.

The previous edition of this title was published as the *Larousse Dictionary of Science and Technology* in 1995. It contains more than 50,000 entries on a wide range of subjects.

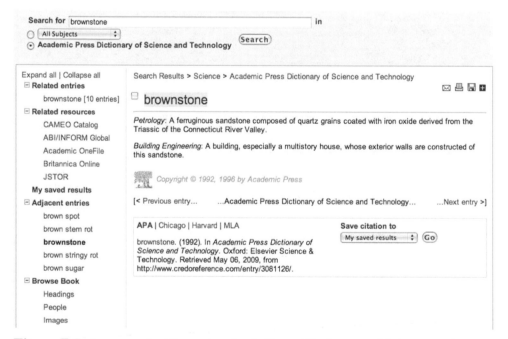

Figure 7.5. A screen shot from *Academic Press Dictionary of Science and Technology.* **Accessed May 2009. Permission granted by Credo Reference.**

3. *McGraw-Hill Dictionary of Scientific and Technical Terms.* 6th ed. New York: McGraw-Hill, 2003. ISBN: 007042313X.

The first edition of this title was published in 1974, partly in response to the need for a specialized dictionary in science for the general reader as well as scholars. This edition includes 110,000 terms and 125,000 definitions together with 3,000 black and white illustrations. Synonyms, acronyms, abbreviations, structural formulas, and math equations are given within the definition. In addition, pronunciation for all terms is included and is unique to this title. Each entry is classed into 1 or more of 104 fields, ranging from general categories such as astronomy and physics to specialized ones such as engineering acoustics and naval architecture where the category abbreviations are inserted in the definitions. The appendix contains information on measurement systems, conversion factors, chemical nomenclature, periodic table of the elements, elementary particles, mathematical signs and symbols, an outline of the classification of living organisms, and brief biographies of Nobel laureates and other individuals after whom scientific terms are named. This is a very comprehensive dictionary in all fields of science and technology. This dictionary is also available through Credo Reference. An interesting feature is the capability of searching in one of three categories: headings, people, or images. Figure 7.6 shows a screen shot of images from *McGraw-Hill Dictionary of Scientific and Technical Terms.*

When the various specialized fields and subfields of science and technology are considered, there are literally hundreds of specialized dictionaries from such diverse areas as biology, chemistry, engineering, mathematics, and physics.

Four representative dictionary titles from the field of chemistry include:

1. *A Dictionary of Chemistry.* 6th ed. Edited by John Daintith. New York: Oxford University Press, 2008. ISBN: 9780199204632.

This updated edition of more than 4,700 entries covers all aspects of chemistry ranging from biochemistry to physical chemistry. In addition, related fields such as forensics, geology, and metallurgy are featured. Additional information includes chronologies and biographies, as well as feature articles and eight appendixes. The text is enhanced with chemical structures, illustrations, and tables. More than 200 entries include recommended Web links found on Oxford University Press's Companion Web Site. Each of these authoritative links provides more detailed information of the relevant term.

2. *Grant and Hackh's Chemical Dictionary.* 5th ed. Completely Revised and Edited by Roger Grant and Claire Grant. New York: McGraw-Hill, 1987. ISBN: 0070240671.

This dictionary covers 55,000 chemical terms and chemical names. It includes synonyms, generic and trade names, as well as physical and chemical properties such as chemical formula, molecular weight,

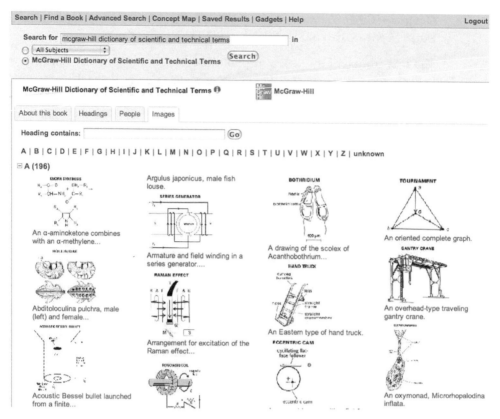

Figure 7.6. A screen shot of images from *McGraw-Hill Dictionary of Scientific and Technical Terms*. Accessed May 2009. Permission granted by Credo Reference and McGraw-Hill Publishing Co.

melting point, solubility, and related chemical data. There is also information from related fields including biology, medicine, mineralogy, and pharmacology.

3. *Hawley's Condensed Chemical Dictionary*. 15th print ed. Revised by Richard J. Lewis Sr. New York: Wiley, 2007. ISBN: 9780471768654.

    Here is an extensive 1,300-page compilation of technical data and descriptive information, particularly from the fields of chemistry, biochemistry, and biology, that covers thousands of chemicals and chemical concepts, pharmaceuticals, trade name products, raw materials, chemical processes, named reactions, equipment, devices, products, and related terminology. It identifies chemical substances by name, physical properties, occurrence, Chemical Abstracts Service (CAS) Registry Number, chemical formula, potential hazards, derivations, synonyms, and applications. There are also abbreviations and numerous cross-references, as well as information on notable chemists and Nobel Prize winners. This title has been a standard in the field for almost a century. It is also available in CD-ROM and as an online resource.

4. *Dictionary of Organic Compounds.* 6th print ed. New York: Chapman & Hall, 1996. ISBN: 0412540908. Available electronically from Chapman & Hall.

In the print edition, the nine-volume set contains concise chemical, structural, and bibliographic information on over 250,000 of the most important organic compounds. These entries have been selected by an international team of experts in the field of modern organic chemistry. Compounds included cover a wide range of commonly used chemicals, synthetic reagents, pesticides, starting materials, and simple fundamental compounds, as well as a selection of natural products, pharmaceuticals, and analytical reagents. Data provided includes names and synonyms, formulas, chemical structures, CAS Registry numbers, physical properties, and key citations from the chemical literature. There are numerous cross-references and three separate indexes, each in its own volume: (1) name index, (2) molecular formula index, and (3) CAS Registry number index. This is an interesting title since it contains characteristics of both a dictionary (an alphabetical arrangement) as well as an encyclopedia (a multivolume set that is comprehensive in the information presented). Figure 7.7 shows a sample page from *Dictionary of Organic Compounds.*

ENCYCLOPEDIAS

If a dictionary is basically a list of words, an encyclopedia usually provides additional and more detailed information. Depending on the publication, information in encyclopedias can range from dictionary-type definitions to extensive essays complete with illustrations and bibliographies. The oldest and most comprehensive encyclopedia to be published is Zedler's *Universal-Lexikon* published in 64 volumes between 1732 and 1754. Small works, called encyclopedias, began to appear from about 1709 onward. The field of science and technology has many encyclopedias ranging from those that cover all areas of science and technology to those that are limited to particular subject areas, such as general biology, biochemistry, genetics, or microbiology within the biological sciences. Similarly, there is specialization in the encyclopedias available within the physical sciences as well as engineering.

Articles in encyclopedias are generally arranged alphabetically and usually include cross-references to other articles. However, some encyclopedias arrange articles topically, within broad subjects, such as astronomy. Science encyclopedias range from a single volume to multivolume sets that may be published all at once or over several years and may include supplements. Some encyclopedias are updated annually by a separate publication, usually referred to as a yearbook. Larger sets of scientific and technical encyclopedias often have an overall index to facilitate locating subjects that may be found in several or many volumes.

As with dictionaries, encyclopedias are available online as well as print and have many added features and capabilities in their electronic form. Along with the standard article and illustrations, e-format contains videos, interactive maps, dynamic

Desyl bromide. *See* α-Bromo-α-phenylacetophenone.

Desyl chloride. *See* α-Chloro-α-phenylacetophenone.

2-Desylpropionic Acid. *See* 3-Benzoyl-2-methyl-3-phenylpropionic Acid.

3-Desylpropionic Acid. *See* 4-Phenyl-4-benzoylbutyric Acid.

Detamide. *See under* *m*-Toluamide.

Dethiobenzylpenicillin (*Desthiopenicillin G*)

$C_{16}H_{20}O_4N_2$ MW 304
Plates from $CHCl_3$ or EtOH.Aq. M.p. 109–11°. pK 3·5 in 8% EtOH.Aq. Hot HCl.Aq. → phenylacetic acid. Anhyd. HCl in dioxan → *d*-valine hydrochloride.
Me ester: $C_{17}H_{22}O_4N_2$. MW 318. Plates from $CHCl_3$–pet. ether. M.p. 108–10°.
Benzylamine salt: plates from MeOH–Et$_2$O. M.p. 149–51° decomp. $[α]_D^{23}$ +9° in H_2O.
N-Benzyl: needles from Me$_2$CO.Aq. M.p. 77–9°. $[α]_D^{25}$ +9·6° in 0·5*N*-EtOH–HCl (initial) → −12·8° (final) in 135 mins. pK (acid) 3·8. *Benzylamine salt*: m.p. 124–7° decomp.

Clarke, Johnson, Robinson, *The Chemistry of Penicillin*, Princeton University Press, 1949.

Dethiobiotin. *See* 6-(5-Methyl-2-oxo-4-imidazolidinyl)-hexanoic Acid.

Dethiopenicillin G. *See* Dethiobenzylpenicillin.

Detigon. *See* Chlophedianol.

3-Deutero-2-bromobutane. *See* 2-Bromobutane-3-*d*.

2-Deuterobutane. *See* Butane-2-*d*.

3-Deutero-2-butanol. *See* 2-Butanol-3-*d*.

Deuterochloroform. *See* Chloroform-*d*.

Deuterohemin. *See under* Deuteroporphyrin IX.

Deuteroporphyrin III

$C_{30}H_{30}O_4N_4$ MW 510
Di-Me ester: $C_{32}H_{34}O_4N_4$. MW 538. M.p. 290°. *Fe salt*: m.p. 285°. *Cu salt*: m.p. 288°.
Fischer, Nussler, *Ann.*, 1931, **491**, 162.

Deuteroporphyrin IX (*"Natural" deuteroporphyrin, copratoporphyrin, pyroporphyrin*)

$C_{30}H_{30}O_4N_4$ MW 510
Found in putrefied blood and in faeces. Cryst. from AcOH–Et$_2$O.
Di-Me ester: $C_{32}H_{34}O_4N$. MW 538. M.p. 223°. *Cu complex*: m.p. 234°.
Di-formyl: m.p. 290°. *Di-Ac*: m.p. 235–6°.
Di-Et ester: m.p. 148°. *Picrolonate*: m.p. 212°.
Flavianate: m.p. 204°.
Styphnate: m.p. 183°.
Picrate: m.p. 240°.
Picrolonate: m.p. 118°.
Flavianate: m.p. 275°.
Fe salt: deuterohemin, copratin, pyratin. Red-brown solid. Conc. H_2SO_4 → violet sol. *Di-Me ester*: m.p. 233°.
Cu salt: m.p. 335°.

Schumm, *Z. physiol. Chem.*, 1928, **176**, 122.
Fischer, Lindner, *Z. physiol. Chem.*, 1926, 161, 17.
Walter, *J. Biol. Chem.*, 1952, **196**, 151.
Chu, Chu, *J. Am. Chem. Soc.*, 1952, **74**, 6276.

Deuteroporphyrin XIII

$C_{30}H_{30}O_4N_4$ MW 510
M.p. 243–243·5°.

Corwin, Krieble, *J. Am. Chem. Soc.*, 1941, **63**, 1829.

Deuticulatol (5 : 7(or 6 : 8)-*Dihydroxy-1-methylphenanthraquinone*)

$C_{15}H_{10}O_4$ MW 254
Constituent of To-Tan-Hwang (*Rumex deuticulata*). Cryst. M.p. 162·5°. Zn dust dist. → 1-methylphenanthrene. Ox. → 3-methylphthalic acid.

Chi *et al.*, *J. Chinese Chem. Soc.*, 1947, **15**, 21.

Figure 7.7. A sample page from *Dictionary of Organic Compounds*, 6th printed. New York: Chapman & Hall, 1996.

time-tables, games, search bars on each page, research organizers, and pointers to Web sites of interest, as well as full-text links to magazine articles on the topic of interest. Encyclopedia software for a personal computer often ranges in cost from $5 to upwards of $50. Computing support is available as a standard feature and is obtainable by phone, chat, and e-mail.

Some of the criteria used to evaluate encyclopedias in science and technology include:

- How easy is it to locate the information you are looking for?
- What is the arrangement and outline of the material?
- What is the scope, depth, and level of detail of the information and is it correct and up-to-date?
- What is the writing style?
- Are there illustrations and other visual aids including tables, charts, graphs, line drawings, photographs, and maps?
- Does the entry include historical information?
- Are there time lines and other features such as sidebar boxes?
- Are there cross-references?
- Is there a bibliography or other suggestions for further reading?
- Is the entry signed and, if so, what is the author's affiliation?

General encyclopedias in science and technology range from a single volume to multivolume sets with annual supplements. Three titles that are representative of the general encyclopedias in science and technology are:

1. *Van Nostrand's Scientific Encyclopedia.* 10th ed. Edited by Glenn D. Considine. Hoboken, NJ: Wiley, 2008. ISBN: 9780471743385. Available electronically from Wiley Interscience.

    The first edition of this title was published in 1938. This three-volume set, although smaller than several multivolume publications, covers all areas of science and technology. The discussions are concise and accessible. The scope ranges from the introductory to the highly technical. In the presentation and discussion of information, there is a progressive development of each topic, beginning with a simple definition and continuing into a more detailed treatment. The suggested readings at the end of each article include both print and Internet references. Detailed time lines and glossaries have been added to some of the larger articles such as artificial intelligence and vision and the eye. Brief biographies of numerous scientists whose work is alluded to in the text are included. For a three-volume set, the numbers are impressive: more than 10,000 entries, approximately 9,500 cross-references, an alphabetical index of more than 100 printed pages, and more than 4,800 illustrations consisting of line drawings, graphs, photographs, and tables. Figure 7.8 shows a screen shot from *Van Nostrand's Scientific Encyclopedia.*

2. *Gale Encyclopedia of Science.* 4th ed. Edited by K. Lee Lerner and Brenda Wilmoth Lerner. Detroit, MI: Thomson Gale, 2008. 6 vols. ISBN: 9781414428772.

This encyclopedia is a one-stop resource written for the nonspecialist. It has great appeal to students in middle school and high school as well as the general adult reader. Topics have been chosen that present basic information in key areas across the science curriculum and provide additional information for those topics that are related to current issues and events. The level of scientific information is midway between that found in an introductory source and that found in a highly technical source. Entries are alphabetically arranged across all volumes in a single sequence. The entries vary in length from short definitions and information in several paragraphs to rather lengthy, detailed presentations on more complex topics and subjects. These longer entries begin with an overview of the subject, followed by a detailed discussion logically arranged under subheadings. A list of key terms is frequently provided where appropriate to define new or unfamiliar terms or concepts. A resources section consisting of books, periodicals, and Web sites accompanies longer entries that are signed

Figure 7.8. A screen shot from *Van Nostrand's Scientific Encyclopedia*, **10th ed. Edited by Glenn D. Considine. Hoboken, NJ: Wiley, 2008. Reprinted with permission of John Wiley & Sons, Inc.**

by an author whose affiliation can be found in the contributors section of each volume. Cross-references direct readers to where information on topics without their own entries can be found. There is a comprehensive general index that covers all topics, illustrations, tables, and individuals mentioned in this set.

3. *McGraw-Hill Encyclopedia of Science & Technology.* 10th ed. New York: McGraw-Hill, 2007. ISBN: 9780071441438.

With 7,100 articles, this encyclopedia provides authoritative, comprehensive information in all fields of science and engineering. This title sets the standard for science encyclopedias. It provides clear discussions on the newest theories and latest research. Many of the entries are new or rewritten, while others have been updated. Articles generally begin with a definition of the topic followed by an overview that then proceeds into more detailed, advanced coverage according to a clear outline and concludes with a bibliography of publications for further study. These individual bibliographic items number almost 25,000. More than 60,000 cross-references permit topics to be further explored and help make connections among topics. There are more than 12,000 illustrations including almost 100 full color plates, and more than 1,400 tables, 900 chemical structures, 2,500 reactions, and 8,500 mathematical equations. Articles are arranged alphabetically, and to find all the information available on a particular subject, there is an analytical index of more than 500 pages. The topical index lists all of the articles pertaining to a particular discipline such as astronomy or microbiology. A comprehensive list of the contributors and their affiliations together with the titles of the articles they wrote is also included in volume 20, the index. New to this edition is a companion Web site which contains materials that complement the printed encyclopedia. Included is information from AccessScience (http://www.accessscience.com/), the online version of the *McGraw-Hill Encyclopedia of Science & Technology* that contains regular updates of articles, images, animations, and videos, as well as commentaries on articles and explorations of topical themes. In addition, AccessScience contains 110,000+ definitions from the *McGraw-Hill Dictionary of Scientific and Technical Terms,* 2,000 biographies from the *Hutchinson Dictionary of Scientific Biography,* as well as the latest news in science and technology from the *Science News* and *ScienceCentral* videos. AccessScience is designed to appeal to students, teachers, and librarians. The *McGraw-Hill Encyclopedia of Science & Technology* is updated annually (between new editions) with the *McGraw-Hill Yearbook of Science & Technology.* The fifth edition of the *McGraw-Hill Concise Encyclopedia of Science & Technology* was published in 2005.

Just as there are hundreds of specialized subject dictionaries in almost every area of science and engineering, there are also hundreds of specialized subject encyclopedias in numerous areas of the biological sciences, the physical sciences, and engineering. Three examples of specialized encyclopedia titles from science and technology are:

1. *Encyclopedia of Biological Chemistry*. Edited by William J. Lennarz and M. Daniel Lane. Boston: Elsevier Academic Press, 2004. ISBN: 0124437109. Available online via Science Direct (http://www.info.sci encedirect.com).

   This four-volume set consists of more than 500 entries that encompass all aspects of biochemistry as well as the extensions of this subject into related fields such as molecular biology, cell biology, genetics, microbiology, and biophysics. Articles are authored by authorities and contain more than 1,300 illustrations including 800 images in four-color and more than 200 tables. Each article begins with a concise definition of the subject that includes general background and term definitions as well as a comprehensive review of the current research in the field. The figures, tables, and other illustrations support and amplify the article text. Each article contains a glossary section that defines key terms used in the article, a bibliography of books and articles for further reading, and a biography of the article's author. The front matter contains an extensive alphabetical list of the contents (i.e., topics such as centromeres, chloroplasts, coenzyme Q, etc.). One section covers the contents by volume and another one the contents by subject area (e.g., bioenergetics). There is an extensive subject index at the end of volume 4. Figure 7.9 shows a sample page from *Encyclopedia of Biological Chemistry*.

2. *Encyclopedia of Chemical Processing*. Edited by Sunggyu Lee. New York: Taylor & Francis, 2006. ISBN: 0824755634. Available online via Informaworld (http://www.informaworld.com/smpp/home~db=all).

   This five volume set is a comprehensive work on chemical processing that consists of nearly 350 articles on current design, engineering, and manufacturing practices within the field. It covers processing novel materials, emerging process technologies and resultant materials, and manufacturing organic and inorganic chemicals. Specific topics include synthesis reactions, properties and characterization of materials, selection of catalysts, reactor design, process flow sheets, energy integration practices, and environmental aspects of chemical plant operation. There is extensive coverage of polymerization and polymer processing as well as advanced materials such as ceramics, nanomaterials, biomaterials, and biomedical materials. This title is published in both print and online formats. Additional articles are added quarterly to the online database. The front matter consists of the contributors and a topical table of contents for each volume. Figure 7.10 shows a screen shot from *Encyclopedia of Chemical Processing*.

Cell Cycle Controls in G_1 and G_0

Wenge Shi and Steven F. Dowdy

University of California, San Diego School of Medicine, La Jolla, California, USA

The cell cycle is the process by which one cell becomes two. Somatic cell division involves cell growth (increase in cellular components, such as ribosomes, membranes, organelles) throughout the cell cycle, faithful replication of its DNA during the S phase of the cell cycle, and precise distribution of DNA between daughter cells during the mitosis (M) phase. In addition, with the exception of early embryonic cell divisions, two gap phases, Gap 1 (G_1) and Gap 2 (G_2), are separated by S phase. As multicellular eukaryote organisms regulate cell division tightly to maintain tissue homeostasis, the most important regulatory decision is made during exiting of the resting state so called the G_0 phase and G_1 phase of the cell cycle before cells become committed to initiate DNA synthesis and complete the cell cycle.

G_1 and G_0 Phase of the Cell Cycle

The majority of cells in adult metazoans are permanently withdrawn from the cell cycle in a terminally differentiated state. Only small numbers of cells, such as hematopoietic and epithelial early progenitor cells, are actively proliferating. Other cells are reversibly withdrawn from the cell cycle and remain in a quiescent stage or the G_0 phase of the cell cycle. Upon proper stimulation, these cells can re-enter the cell cycle. For example, highly differentiated hepatocytes in adult liver are present in the G_0 resting phase and rarely replicate normally. However, in response to acute liver injury or distress, these G_0 hepatocytes can be stimulated to re-enter the cell cycle and regenerate the liver. Similarly, primary peripheral blood lymphocytes (T and B cells), which are present in the G_0 phase, can re-enter the cell cycle and start clonal expansion when presented with the appropriate antigen. These examples serve to demonstrate that certain cell types may enter and exit the cell cycle pending the appropriate stimulus, whereas the vast majority of cells in a matured metazoan have permanently exited the cell cycle.

The most important decision for cell cycle progression is made during the G_1 phase. In G_0 phase, cells respond to extracellular signals by entering the G_1 phase of the cell cycle. However, prior to transiting into the late G_1 phase, they must traverse the growth factor-dependent restriction point. This is a critical regulatory phase of the cell cycle where a cell becomes committed to enter S phase and finish the remaining cell cycle. In contrast to normal cells, tumor cells are often less growth factor-dependent and fail to respond to growth inhibitory signals. Consequently, tumor cells select for genetic mutations that disrupt the important decisions performed at the restriction point and this is, in fact, one of the hallmarks of cancer.

Regulators of G_0 Exit and G_1–S Progression

One of the key negative regulators of the restriction point and early G_1 to G_0 cell cycle exit is the retinoblastoma tumor suppressor protein (pRb) and two closely related family members, p107 and p130. Murine embryonic fibroblasts (MEF) deficient for all three pocket proteins fail to respond to G_1 phase arrest signals following contact inhibition, serum starvation, or DNA damage. The antiproliferation function of the pocket proteins depends, at least in part, on their interaction with E2F/DP transcriptional factors, which regulate the expression of key genes required for DNA synthesis, DNA repair, DNA-damage checkpoint, apoptosis, and mitosis. Direct interactions of pRb family members with E2F/DP complexes and the recruitment of chromatin-modifying enzyme complexes, such as histone deacetylases (HDAC), polycomb group proteins, the SWI/SNF complex, and histone methyl transferases, to E2F-reponsive genes results in inhibition of target gene expression.

Distinct pRb family member–E2F repressor complexes exist in different cell cycle phases. As an example, during the G_0 phase, p130 and p107 interact with E2F4 and E2F5, whereas pRb remains unbound. E2F4 and E2F5 are expressed constitutively and are involved in E2F-dependent gene repression during cell cycle exit and terminal differentiation. It is also reported that E2F6, the sole E2F family protein that does not interact with the pocket proteins, mediates gene repression with polycomb group proteins in the G_0 phase. Further studies are

Figure 7.9. A sample page from *Encyclopedia of Biological Chemistry*. Edited by William J. Lennarz and M. Daniel Lane. Boston: Elsevier Academic Press, 2004.

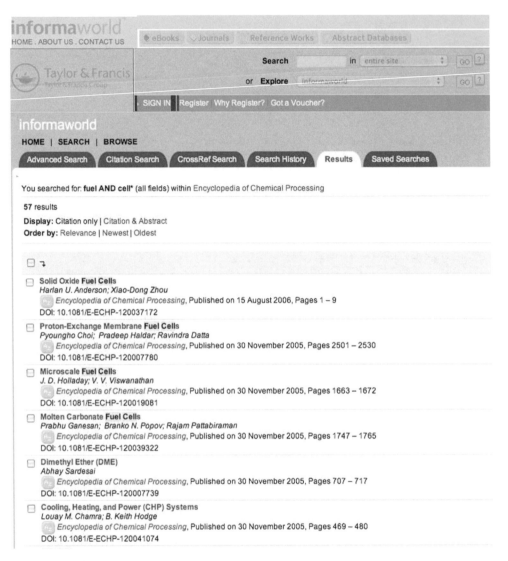

Figure 7.10. A screen shot from *Encyclopedia of Chemical Processing*. Edited by Sunggyu Lee. New York: Taylor & Francis, 2006. Reprinted with permission of John Wiley & Sons, Inc.

3. *Kirk-Othmer Encyclopedia of Chemical Technology.* 5th ed. Hoboken, NJ: Wiley-Interscience, 2004. ISBN: 0471488100. Available online through Wiley Interscience (http:// www.interscience.wiley.com).

   Widely known as "the Bible of Chemical Technology," the *Kirk-Othmer Encyclopedia of Chemical Technology* is probably the most indispensable tool to chemists and chemical engineers with the exception of Chemical Abstracts. The first edition of this title was published in 1949. The fifth edition, consisting of 27 volumes, builds upon the solid foundation of the previous editions, which have proven to be

Home / Chemistry / Industrial Chemistry

Kirk-Othmer Encyclopedia of Chemical Technology
Copyright © 2001 by John Wiley & Sons, Inc.
Last updated: 23 Apr 2009

Reference Work Home I What's New I For Authors I How to Order I Sample Content

Search Results

There are **200** results for: "**pesticide** within **Kirk-Othmer Encyclopedia of Chemical Technology**"

View: 1 - 25 I 26 - 50 I 51 - 75 I 76 - 100 I 101 - 125 I Next >

Article Information

Food Safety, Chemical Contaminants and Toxins
Carl K. Winter
Published online: 14 September, 2007
Abstract I Full Text: HTML I PDF (104K)

Soil Chemistry of Pesticides
Philip C. Kearney, Daniel R. Shelton, William C. Koskinen
Published online: 04 December, 2000
Abstract I Full Text: HTML I PDF (413K)

Pesticides
Marguerite L. Leng, Updated by Staff
Published online: 23 September, 2005
Abstract I Full Text: HTML I PDF (233K)

Controlled Release Technology, Agricultural
Richard M. Wilkins
Published online: 18 June, 2004
Abstract I Full Text: HTML I PDF (211K)

Herbicides
Jack R. Plimmer, Judith M. Bradow, Christopher P. Dionigi, Richard M. Johnson, Suhad Wojkowski
Published online: 17 September, 2004
Abstract I Full Text: HTML I PDF (381K)

Figure 7.11. A screen shot from *Kirk-Othmer Encyclopedia of Chemical Technology*, 5th ed. Hoboken, NJ: Wiley-Interscience, 2004. Reprinted with permission of John Wiley & Sons, Inc.

a mainstay for chemists and engineers at academic, industrial, and government institutions. *Kirk-Othmer* has been described as the best reference of its kind. This set contains more than 1,000 articles on a wide range of topics including chemicals and chemical substances, properties, manufacturing, and uses, as well as information on industrial processes, unit operations in chemical engineering, and numerous additional subjects related to the field. Current research, new and emerging technologies, economic aspects, and environmental and health concerns are also included. Attorneys and consultants frequently consult this title for information on the state-of-the-art in chemical technology for litigation and patent support. Interestingly, various products, processes, and concepts once believed obsolete are again receiving more attention because of changing economics, raw materials shortages, or environmental concerns. This is a principal reason why older editions of this title can maintain a surprisingly long, useful life span. *Kirk-Othmer* is comprehensive, authoritative, accessible, and lucid, providing expert information by the leading authorities in academia and industry. It is one of the most reviewed titles in all of science and technology. Monthly updates keep the online version on the cutting edge of chemical technology. Previous versions of revised articles are archived for posterity. Figure 7.11 shows a screen shot from *Kirk-Othmer Encyclopedia of Chemical Technology*.

Dictionaries and encyclopedias provide specific information usually in alphabetical order, which makes the information easy to locate; other resources can also be helpful. The next chapter covers handbooks, tables, and manuals.

Chapter

8

Handbooks, Tables, and Manuals

HANDBOOKS

Handbooks are reference books in which data, both numerical and textual, is gathered from a variety of sources, organized, and presented for quick and easy use and consultation. Handbooks cover a wide range of subjects and topics and are compendiums of information in a particular field or deal with a specialized technique, such as thin layer chromatography. The wide variety of sources housing the assembled data includes journals, reviews, encyclopedias, and other multivolume titles. Presentation of the data is usually in the form of charts, tables, and lists. Some handbooks are especially useful in graduate and undergraduate science and engineering courses. This includes such titles with equations and formulas from a particular field such as physics. In order for the quantitative data to be accessed and retrieved quickly, indexes that are both extensive and detailed are required. Frequent updating with recent information, together with appropriate references, is also necessary. Handbooks are common in both the basic sciences such as biology and chemistry as well as the applied sciences, particularly engineering.

Textual information in handbooks frequently provides a quick review of a topic together with numerical values from tables within a particular subject field. Some handbooks are general and cover a discipline such as chemistry or physics, while others are more specialized and devoted to a subject area such as ergonomics or acoustics. Most handbooks are a one-volume publication, but there are many titles that are multivolume sets. In some instances, the information is continuously presented, for example, alphabetically in several volumes or each volume is devoted to a particular aspect of the subject. Finally, The Web-based application Knovel is a database that combines the capability of searching data and inquiries in full-text handbooks (including information located in graphs and tables) with interactive analysis tools designed to enable the researcher to manipulate, analyze, and export the data to spreadsheets (see the appendix to this volume).

TABLES

Tables are usually reference publications consisting of numerical data, particularly from such fields as mathematics and statistics. Such information may be published as a separate chapter or an appendix in a work such as a textbook or gathered together with other tables and published as a collection. In addition to mathematics and statistics, other subject data from such fields as astronomy, meteorology, and engineering (e.g., steam tables) is frequently published in tabular format. Data such as boiling points, melting points, solubility, density, and related information easily lends itself to presentation in a table.

MANUALS

Manuals are a type of reference publication that provide detailed explanations of how to do something, how a piece of equipment works or is operated, or how a process or procedure is carried out. They are often referred to as instruction or guide books. More text in the form of explanatory or background material, rather than tables and related data, is usually found in such works. Manuals usually accompany new equipment and scientific instruments and are frequently necessary to assemble and operate the new purchase. Also, manuals are often consulted to troubleshoot a malfunctioning instrument. Detailed textual explanations and diagrams of parts and components are usually present in manuals. Most manuals of this type are produced by the manufacturers of the instruments or equipment. However, other manuals are subject specific and range from identifying bacteria to preparing laboratory solutions and related materials.

An important reference title that is particularly useful for identifying handbooks and tables in all fields of science and technology is *Handbooks and Tables in Science and Technology* by Russell H. Powell. The third edition was published in 1994 by Oryx. It lists almost 3,700 handbooks and tables in all fields of science and technology. Coverage is limited to titles containing hard data that can be used to answer reference questions, and it includes government publications and other research reports. Arrangement is alphabetical by title and complete bibliographic information is included. In addition, most entries have annotations that describe the content of a particular title. Although dated, it is still useful as both a collection development and evaluation tool and in ascertaining whether or not a handbook has been published (at least through the mid-1990s) that deals with a specific field or subject.

One important example of a handbook is:

International Critical Tables of Numerical Data, Physics, Chemistry and Technology. National Research Council. New York: McGraw-Hill, 1926–1930. National Library: 04230030R, LCCN: 26–10495.

This seven-volume reference set is one of the oldest, largest, and most frequently consulted science titles and continues to be an important source for information in various fields of science, engineering, and technology. It is divided into 300 sections and contains "critical" data, that is, data that is considered to be the best value in the opinion of a specialist that contributed the information.

Other examples of handbooks, primarily from the field of chemistry, include:

1. *CRC Handbook of Basic Tables for Chemical Analysis.* 3rd ed. Thomas J. Bruno and Paris D. N. Svoronos. Boca Raton, FL: CRC Press, 2010. ISBN: 9781420080421.

   This reference title includes tables that are classified by analysis method and range from electrophoresis to ultracentrifugation. Useful in all areas of chemistry, especially analytical chemistry, new tables cover radiation safety and units as well as information on laboratory glassware usage.

2. *CRC Handbook of Chemistry and Physics.* 90th ed. Edited by David R. Lide. Boca Raton, FL: CRC Press, 2009. ISBN: 9781420090840.

   This is one of the most important reference titles that provides basic information and data in the fields of chemistry and physics. It has complete coverage of virtually every element, numerous physical formulas, and many mathematical tables. There are 19 sections ranging from basic constants, units, and conversion factors to properties of polymers together with an extensive index. This title is available in print format, CD-ROM, and online.

3. *CRC Handbook of Data on Organic Compounds.* 3rd ed. Edited by David R. Lide and G. W. A. Milne. Boca Raton, FL: CRC Press, 1993. ISBN: 9780849304453.

   This seven-volume reference set contains current, accurate, and clearly presented chemical, physical, and spectral data for more than 27,000 organic compounds.

4. *Chemical Technicians' Ready Reference Handbook.* 4th ed. Gershon Shugar and Jack Ballinger. New York, McGraw-Hill, 1996. ISBN: 9780070571860.

   This reference title deals with laboratory safety, analytical procedures, and instrumentation techniques in the laboratory environment. It contains an extensive index and the contents are a good review for the experienced technician or useful in training new or inexperienced laboratory personnel.

5. *The Chemist's Companion: A Handbook of Practical Data, Techniques, and References.* Arnold J. Gordon and Richard A. Ford. New York: Wiley, 1972. ISBN: 9780471315902.

   This volume contains a wide variety of practical, everyday information frequently required by chemists but not usually found together in one publication. It covers physical, chemical, and mechanical properties of substances and numerous laboratory recipes. Much of the information is in tables and referenced.

6. *Lange's Handbook of Chemistry.* 16th ed. Edited by James Speight. New York: McGraw-Hill, 2005. ISBN: 0071432205.

This is an important and major compilation of facts, data, tabular material, experimental findings, and other information in all areas of chemistry. It is organized into four sections: general information and conversion tables, spectroscopy, inorganic chemistry, and organic chemistry. Included are equations that permit calculations of values such as temperature and pressure as well as practical laboratory information.

7. *The Merck Index: An Encyclopedia of Chemicals, Drugs, and Biologicals.* 14th ed. Edited by Maryadele J. O'Neil. Whitehouse Station, NJ: Merck, 2006. ISBN: 9780911910001.

This is one of the most important reference titles that is available for the chemist. It provides detailed information for more than 10,000 common chemicals in disciplines as diverse as agriculture and medicine. It covers chemical, common, and generic names and offers more than 8,500 chemical structures, 15,000 trademarks, and over 12,000 Chemical Abstracts Service Registry Numbers.

8. *Perry's Chemical Engineers' Handbook.* 8th ed. Don W. Green, Editor-in-Chief. New York: McGraw-Hill, 2008. ISBN: 9780071422949.

This is the standard reference title in chemical engineering and covers important data and information on every aspect of chemical engineering from fundamental principles to chemical processes and equipment, including computer applications. There are comprehensive tables and charts for unit conversion and a major section on physical and chemical data.

9. *Sax's Dangerous Properties of Industrial Materials.* 11th ed. Edited by Richard J. Lewis Sr. and N. Irving Sax. Hoboken, NJ: Wiley-Interscience, 2011. ISBN: 0471476625.

This three-volume set is a major compendium of information about the dangerous properties of more than 26,000 industrial materials and related compounds. In addition to toxicological data, it also covers reactivity, fire, explosive potential, and regulatory information from OSHA and related agencies.

10. *Standard Methods of Chemical Analysis.* 6th ed. N. Howell Furman et al. Malabar, FL: Krieger Reprint, 1975. ISBN: 9780882753409.

This three-volume set includes the major methods of chemical analysis that are used frequently and are accurate. In addition to the elements and industrial and natural products, instrumental and noninstrumental methods are included.

This chapter has covered handbooks, tables, and manuals, reference tools that are critical to the work of an engineer. The next chapter will explain intellectual property rights. Maintaining ownership of ideas through patents, trademarks, and copyright are at the heart of every scientist's and engineer's work when someone else's use of another individuals creativity can be costly to the person with the original idea.

Chapter

9

Intellectual Property: Patents, Trademarks, and Copyright

According to the World Intellectual Property Organization (WIPO), intellectual property is "a power tool for economic development and wealth creation that is not yet being used to optimal effect in all countries, particularly in the developing world" (Kamil 2003, 5).

Your house, your car, and your civil rights are controlled and protected by laws. Your ideas are not, though; not, that is, until you communicate them. Ideas are intellectual property (IP) and only when they are communicated, translated into writings, or realized in tangible objects do legal rights and obligations arise.

No one is obligated to communicate ideas; however, keeping them to yourself does not serve humankind. Patenting an idea is a way to protect your intellectual property and to contribute to the common good. The authors of the U.S. Constitution understood this and Article I, Section 8, Clause 8, of the U.S. Constitution, known as the Copyright Clause, the Copyright and Patent Clause (or Patent and Copyright Clause), the Intellectual Property Clause and the Progress Clause, empowers the U.S. Congress

> to promote the Progress of Science and the Useful Arts, by securing for limited Times to Authors and Inventors the exclusive Right to their respective Writings and Discoveries.

PATENTS

In the United States the first modern patent act was passed in 1790 and the country's first patent was granted to Oliver Evans for his automatic gristmill (Patent 2009). Other nations followed the U.S. lead, with France instituting patent law the following year. Patents in the U.S. system did not receive numbers until 1836; in 2009 the patent numbering system was up to over seven million patents (Patent 2009).

As early as the 15th century, patents were awarded in Europe by rulers of the various nations to individuals they wished to curry favor with rather than in reward for creativity and for the benefit of humankind. Galileo applied for a Venetian patent in 1593 for an agriculture water supply pump; his application was examined and he was given a 20-year exclusive right to his invention with a fine of 200 ducats to be imposed onto any infringer of his granted right (Gordon and Cookfair 2000). Today the World Intellectual Property Organization (WIPO), a specialized agency of the United Nations dedicated to developing a balanced and accessible international intellectual property system, has over 184 member states who assist with promoting the protection of IP throughout the world. Based in Geneva, Switzerland, its diverse staff represents 90 percent of the world's countries holding expertise in IP law, economics, public policy, and information technology (IT) (Patent 2009).

In *Scientific American*'s book on inventions and discoveries, Carlisle divides the milestones in ingenuity into five distinct time periods and provides examples in each category (Carlisle 2004).

- *The Ancient World 8000 b.c. to a.d. 300 and Middle Ages through 1599:* Technical progress was made in developing tools, materials, appliances, fixtures, and methods and procedures implemented by farmers, animal herders, cooks, tailors, healers, and builders. Rather than giving credit to one individual for these inventions, the credit is given to humankind in general for these sweeping developments.

  - Examples: agriculture, aqueducts, board games, ice cream, plumbing, the wheel, wine, decimal numbers, sailing vessels, and the mathematical concept of zero.

- *Age of Scientific Revolution 1600–1790:* This is the era of the named single inventor with most achievements attributed to a relatively small handful of people. The key type of instruments invented were those that enabled curious minds to look more closely at the natural universe or to measure natural phenomena. Did the tools produce the discoveries or did scientific curiosity drive the development of the tools?

  - Examples: the telescope and microscope, the mercury barometer and thermometer, the balance spring, and the pendulum clock.

- *Industrial Revolution 1791–1890:* Principles that governed machines were thought through; newly discovered principles were put in place making it possible to build better machines and to reduce principles to mathematical formulas.

  - Examples: interchangeable parts, steam engines, the electric telegraph, transcontinental railroad, the phonograph, lightweight cameras, the typewriter, and news of the first automobiles.

- *The Electrical Age 1891–1934:* Introduction of new consumer items that changed the everyday life of the average person. Change was rapid and life was more comfortable than in previous times but paced much faster.

- Examples: the highlight of this time period was the automobile (a novelty in 1895 but an invention that enjoyed widespread use by 1910), safety razors, aspirin, thermos bottles, electric blankets, breakfast foods, cellophane, rayon, X-rays, and the zipper.

- *Atomic and Electronic Age 1935–present:* During this current period we constantly witness multiple cases of science producing technologies and technologies leading to further scientific advance. This time period has had the major advancement of the discovery of nuclear power and its use in creating weaponry. Many scientific and engineering fields advanced during the world wars and beyond. With the rapid advancement of technology came new concerns about their social consequences.

  - Examples: turbojet engines, computers, DDT, genetic engineering, and the Internet.

Taken together this is an impressive history of the milestones of human ingenuity. By viewing invention through the eyes of our human history on earth, one can see how technology and science work together to improve living for all humankind.

Patent literature is one of the most interesting and largest bodies of technical information. Just imagine: everything ever made by a human being has been patented at one time or another. As we study early humankind's development, we witness the ability of humans to invent and discover ways to survive and to improve upon their condition as they struggle to adapt to their environment. Some of the most marvelous human developments on earth were conceived and built by ancient civilizations as early as 8000 B.C. and the innovation and resourcefulness of humankind continues at full speed in the present time.

WHY SEARCH PATENTS?

Information professionals can provide invaluable information to users on the technology of an invention by searching the patent literature. As reported by the European Patent Office in 2009, over 70 percent of the information in the patent literature is not available anywhere else. Locating journal articles, technical reports, and marketing information is vital for a good industry search, but patent searching is the key to unlocking the functionality and claims of the invention. This is a compelling reason to remember to check patent documents when seeking information of a technical, business, or legal nature.

On the technical side the patent is laid out in sections (more detail to follow in this chapter), and one can think of it as being similar to a cooking recipe; all the ingredients and the methodology for creating the dish are supplied. For a patent to be valid it should enable any person skilled in that particular area to reproduce the invention. All patent documents follow a universal bibliographic format encompassing more than 50 data fields that represent technical or strategic information, all internationally adopted, and falling into over 100,000 subdivisions. These codes are known as INID Codes and appear on the front pages of all patents internationally. For example, all patents have data fields signified by a number (in parenthesis or brackets); the number 10 code holds the patent number, the 45 data field code is always the date of the patent, the 75 data field code is the inventor's name, and so on. Having globally accepted bibliographic practices makes searching across borders and foreign languages seamless. As of January 2010, the United States Patent and Trademark Office (USPTO) has granted over 7,658,800 patents.

From a business and competitive intelligence perspective, often the earliest sign of a company's investment and protection of a new invention is discovered by searching a patent application database. Patent applications are made public 18 months after the first filing date, and this is often the first official indication of a company's otherwise private information. Data from the patent application assists in learning who the company may be collaborating with, such as other companies or research centers in universities. Fees for applying and maintaining patents for their potential 20-year life span are costly and so it follows that most patents currently are paid for by companies and institutions rather than by individual inventors.

From a legal perspective, patents are sometimes infringed upon by others, and if proven, the parties in question can be sued. In this case a validity search is conducted and the claims for prior art inventions are searched. A patent is a legal document that was disclosed at the application stage and vetted by patent experts in the patent granting office. This form of law seeks evidence that would provide the answer of who came up with the original invention and protect their legal rights. Additional information about patents is given later in this chapter.

Trademarks

It is easy to confuse trademarks, service marks, copyright, and patents. These forms of intellectual property all fall under the umbrella of IP and are unique in their form and function. A *trademark* is a word, phrase, symbol, design, or a combination of these that identifies and distinguishes the source of the goods or services of one party from those of others. Trademarks are generally displayed on a product or its packaging or displayed on physical storefront property or delivery vehicles. Trademark protection is usually sought by those wishing to sell products and to create a signature brand that signifies to the public that their product is unique and recognizable and is the source for the product. The owner of a trademark has the right to prevent others from using a similar or matching mark on their product(s) or service(s). To be able to display and use the ® symbol to designate a registered trademark, one must have applied for and been granted federal registration. As long as the mark is used in commerce it retains its registration protection. See Figure 9.1 for an example of a word mark.

A *service mark* is the same as a trademark except that it identifies and distinguishes the source of a service rather than a product. Service marks are used to advertise and brand a service and often appear on printed brochures, sides of service vehicles, on Web sites for the service, and on correspondence. A service mark can be

Figure 9.1. An example of a mark image from the USPTO database. Word Mark: PENGUIN'S PARADISE DESSERT BAR. Accessed September 2009.

a very powerful marketing tool. Do you recognize the service mark in Figure 9.2? It belongs to the Mayflower Transit LLC, the moving company, and has been in service since 1948. The company continues to renew it mark registration and therefore it is currently categorized as a "live" mark.

Copyright

This form of protection is provided to authors of "original works," which include literary, dramatic, musical, and artistic works, and covers the "expression of ideas" not the idea itself. It provides an author with certain exclusive rights for a limited time. Those rights include the right to reproduce the copyrighted work; to prepare derivative works; to distribute copies; and to perform or display the work publicly. Rights begin at the time the work is first created or fixed in a tangible medium of expression from which that work can be communicated or reproduced. Examples of tangible forms include a piece of music recorded on tape, a literary work written down, or a motion picture captured on film. Copyrights can be registered by filing in the U.S. Copyright Office through the application process. Copies of the work along with the payment of a fee are required. A copyright symbol is generally placed on the item and looks like this: ©.

Figure 9.2. An example of a service mark. Accessed September 2009.

WHAT IS A PATENT?

A patent for an invention is a grant of a *property right* by the government to the inventor acting through the Patent and Trademark Office. In the United States the term of a patent is 20 years from the date an inventor first applies for the patent and that term is subject to the payment of scheduled maintenance fees (at years 3.5, 7.5, and 11.5). A patent is a *contract* that gives the inventor "the right to exclude others from making, using, or selling" the invention during a limited time period. What is granted is not the right to make, use, or sell; but rather it is the right to exclude others from doing so. The inventor must provide a full public disclosure of information about the invention and this is done through the application process. We will discuss the parts of a patent later in this chapter.

What Can Be Patented?

Any process, machine, manufacture, or composition of matter can be patented. These classes of subject matter taken together include practically everything made by a human being and the process of making them. A "process" is defined as an industrial or technical process, act, or method that is considered patentable. A "machine" is self-defining. The term "manufacture" refers to articles that are made and includes all manufactured articles. The "composition of matter" refers to chemical structures, their composition as well as including mixtures of ingredients, and the creation of new chemical compounds.

Criteria for Obtaining a Patent

Inventions must meet three criteria to be patentable:

1.   It must be new.
2.   It must be useful.
3.   It must be non-obvious (novel) to a person in that field.

"New," as defined in patent law, allows an invention to be patentable if it was not known, used by others, or described in a printed publication in the patenting country; was not already patented, in use, or described in a printed publication in another country; or was not already in public use or on sale one year prior to the application for a patent in the United States. "Useful" refers to the operativeness/viability of the invention: Will it perform as intended and described and does it serve a useful purpose? The "non-obvious" criterion is judged by whether the subject matter to be patented is sufficiently different from what has been used or described previously in the patent art and that it could be said to be non-obvious to a person having ordinary skill in the technology of the invention.

TYPES OF PATENTS: UTILITY PATENTS, DESIGN PATENTS, AND PLANT PATENTS

The United States patent system has six types of patent documents. The majority of patent applications occur within three types: utility, design, and plant patents.

Utility Patent

Utility patents are the most common form of patent type and cover inventions in the fields of mechanical, electrical, chemical, software, and methods of doing business. Typically these are the inventions we are most familiar with and that make our living easier. Utility patents include inventions such as the computer I am using to type this book, the toothbrush you use each day, and the light switch you turn on and off as necessary for safety and sight. By patenting your invention you are protecting the way in which the invention is made, how it will be used, or how it functions. An easy way to view utility patents is to think of them as "everyday gadgets" that enrich our lives. See Figure 9.3 for an example of a utility patent.

Design Patent

This type of patent covers only the ornamental appearance of an article of manufacture and therefore gives narrow protection to the inventor. Simply put, the design patent

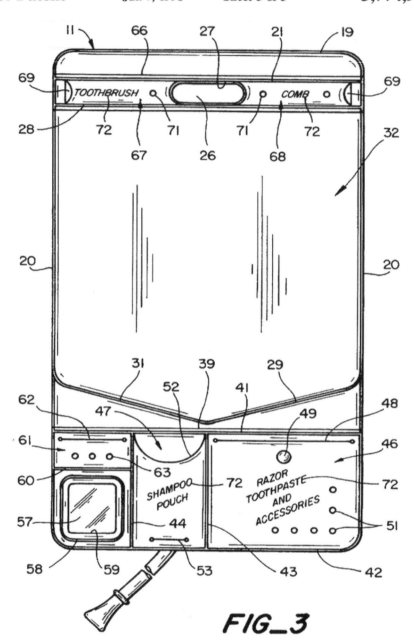

FIG_3

Figure 9.3. Utility Patent US 5,774,908, *Outdoor Shower and Accessory Holder*, from the USPTO Web site. Accessed August 2009.

protects the way an invention looks, not how it functions. As we can see in Figure 9.4, the electronic input device (commonly known as a "mouse") may be very familiar to us and is manufactured by a variety of companies. However, each company produces a "mouse" that ornamentally looks somewhat different from the one in our figure that is protected by the holder of this design patent. Corporations that manufacture these devices are best protected by simple design claims in the patenting process, as it is easy to infringe upon the "original design" to create "knock-offs" that may sell for far less than the original design yet claim to be the original protected design.

Plant Patent

This category of patent is the smallest general category and is utilized mainly by plant breeders and agriculturists. This type of patent is issued for newly invented strains of asexually reproducing plants and does not apply to plants in the wild (natural world).

U.S. Patent   Aug. 11, 2009   Sheet 1 of 10   **US D598,022 S**

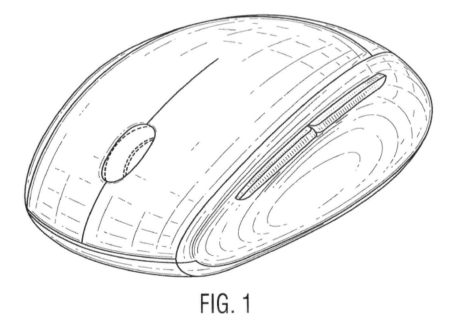

FIG. 1

Figure 9.4. Design Patent US D598,022 S, *Electronic Input Device,* **from the USPTO Web site. Accessed August 2009.**

Figure 9.5. Plant Patent US PP19,876 P2, *Rosa Hybrida* Shrub named "AUSBERNARD," from the USPTO Web site. Accessed August 2009.

Asexually created plants are not generated by seeds but rather by the grafting or rooting of cuttings from existing plants. Plant patents are in force for 20 years from the date of patent application filing and are not subject to the payment of maintenance fees. Figure 9.5 shows an example of a plant patent.

The other three types of patents are reissue, defensive publication, and statutory invention registration patents. For more information on these types of patents, see the United States Code Title 35—Patents.

PARTS OF A PATENT

The parts of a patent include front page, specification section, claims section, and drawing section. These are described below.

Front Page

The front page of a patent contains all the bibliographic information required by the applicant when filling out the application data sheet. Specific types of information include applicant data, correspondence information, dates of submission, inventor name(s), U.S. and international classification codes, the field of classification searching done by the applicant, references cited within the U.S. patent documents, other publications discovered during the prior art search, the patent office examiner name(s), an abstract, and a drawing. Having this abbreviated information on the first page can assist the user of the patent literature during the discovery and search stages. It can also reveal how long the examiner process took for the patent application to journey from the date it is sent to the USPTO offices to the final patent granting stage. In the case of the patent in Figure 9.6, it was filed on January 4, 2000, and granted on September 25, 2001, a period of approximately 21 months. Patents can range in length from 1–2 pages to hundreds of pages.

Specification Section

This section includes a written description of the invention and must be so clear as to enable any person skilled in the technological area of the patent to be able to make and use an exact replica of the invention described. The description must also explain improvements to the invention if it is based on prior art and must explain the mode of operation wherever applicable. See Figure 9.7 for an example of a specification section patent.

Claims Section

Within the specification section one must state all claim or claims distinctly describing the subject matter that the applicant regards as the invention. As shown in Figure 9.8, more than one claim can be provided as long as it is different from others already described.

Drawing Section

Drawings of an invention are required by law in the patent application process, especially if the nature of the invention requires a drawing to understand it. Drawings must show every feature of the invention as laid out in the claims, they must follow the form for writing claims as explained in the application process, and they must be effortlessly understood by readers of the patent description. In the utility or design paper application the applicant is permitted to present drawings in either black ink or color; if color is necessary to disclose the medium of the invention, a fee is imposed. Color drawings are not permitted in international applications. Figure 9.9 shows an example of a drawing section patent.

THE SEVEN STEPS TO SEARCHING THE PATENT LITERATURE

Whether you have access to fee-based patent database search services (e.g., DIALOG, LEXIS-NEXIS, STN, etc.) or must rely on free search sites on the Internet, you will want to learn about the seven critical steps to follow when working with the patent

US006293874B1

(12) **United States Patent**
Armstrong

(10) **Patent No.:** **US 6,293,874 B1**
(45) **Date of Patent:** **Sep. 25, 2001**

(54) **USER-OPERATED AMUSEMENT APPARATUS FOR KICKING THE USER'S BUTTOCKS**

(76) Inventor: **Joe W. Armstrong**, 306 Kingston St., Lenoir, TN (US) 37771-2408

(*) Notice: Subject to any disclaimer, the term of this patent is extended or adjusted under 35 U.S.C. 154(b) by 0 days.

(21) Appl. No.: **09/477,175**

(22) Filed: **Jan. 4, 2000**

(51) Int. Cl.^{7} **A63H 37/00**
(52) U.S. Cl. **472/51**; 472/55
(58) Field of Search 472/51, 55, 137; 482/51, 72, 148

(56) **References Cited**

U.S. PATENT DOCUMENTS

654,611	7/1900	De Moulin .
920,837	5/1909	De Moulin .
953,411	3/1910	De Moulin .
966,935	8/1910	Mamaux .
976,851	11/1910	De Moulin .
1,175,372	3/1916	Newcomb .
4,457,100 *	7/1984	Nightingale 446/333
5,785,601 *	7/1998	Kubesheski et al. 472/135

* cited by examiner

Primary Examiner—Joe H. Cheng
Assistant Examiner—Kim T. Nguyen
(74) Attorney, Agent, or Firm—Pitts & Brittian, P.C.

(57) **ABSTRACT**

An amusement apparatus including a user-operated and controlled apparatus for self-infliction of repetitive blows to the user's buttocks by a plurality of elongated arms bearing flexible extensions that rotate under the user's control. The apparatus includes a platform foldable at a mid-section, having first post and second upstanding posts detachably mounted thereon. The first post is provided with a crank positioned at a height thereon which requires the user to bend forward toward the first post while grasping the crank with both hands, to prominently present his buttocks toward the second post. The second post is provided with a plurality of rotating arms detachably mounted thereon, with a central axis of the rotating arms positioned at a height generally level with the user's buttocks. The elongated arms are propelled by the user's movement of the crank, which is operatively connected by a drive train to the central axis of the rotating arms. As the user rotates the crank, the user's buttocks are paddled by flexible shoes located on each outboard end of the elongated arms to provide amusement to the user and viewers of the paddling. The amusement apparatus is foldable into a self-contained package for storage or shipping.

14 Claims, 7 Drawing Sheets

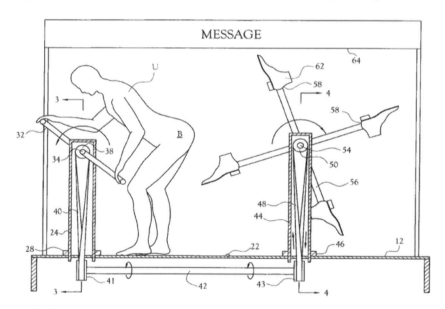

Figure 9.6. The front page of a utility patent, US 6,293,874 B1. Accessed September 2009.

USER-OPERATED AMUSEMENT APPARATUS FOR KICKING THE USER'S BUTTOCKS

CROSS-REFERENCE TO RELATED APPLICATIONS

Not Applicable.

STATEMENT REGARDING FEDERALLY SPONSORED RESEARCH OR DEVELOPMENT

Not Applicable.

BACKGROUND OF INVENTION

1. Field of Invention

The invention relates to a user-operated amusement apparatus. More specifically, the invention relates to an amusement apparatus including a user-operated and controlled plurality of rotating members for self-kicking the user's buttocks.

2. Description of the Related Art

Prior art devices include individual spanking devices that must be reloaded or reset after each individual spanking action. Typical prior art devices provide a paddle that can pivot once, upon being triggered, to spank a hand or buttocks of the user. In U.S. Pat. No. 920,837, issued to De Moulin, a device is disclosed for lifting and spanking of the user for secret society initiation ceremonies. The device includes a trick lifting machine having a spring member, lever, and manual actuation for triggering the paddle release. The actuation by a user releases the spring member, pivoting the paddle, and striking a user straddled over the device. The spring and paddle are reset after each actuation.

This and other known devices of the type, disclose spring activated, individual paddle actions that must be reset after each contact with the user, with associated repositioning of the user in a posture to accept the next individual paddle action.

Therefore, it is an object of the present invention to provide an amusement apparatus which is user-operated and controlled, and is designed to inflict repetitive blows on the user without resetting of the apparatus and/or repositioning of the user between blows.

It is another object of the present invention to provide an amusement apparatus having a user controlled crank regulating the frequency and force of the blows inflicted upon the user's buttocks.

It is another object of the present invention to provide an amusement apparatus for self-inflicting repetitive blows to a user and which is foldable into a self-containing package for storage or shipping.

BRIEF SUMMARY OF INVENTION

An amusement apparatus that includes a user-operated and controlled device for self-infliction of repetitive blows to the user's buttocks including a plurality of rotating arms bearing flexible extensions for self-paddling the user's buttocks. The amusement apparatus includes a platform having a first upstanding post detachably positioned near a first end of the platform, and having a second upstanding post detachably positioned near a second end of the platform. The first post is provided with a crank suitable to be grasped by the user's hands for rotation of the crank at a speed and with a force determined by the user. The second post includes a top end onto which there is mounted a plurality of elongated

arms, each of which includes an outboard end having a pliable paddle, a flexible shoe, or the like mounted thereon. The plurality of arms are rotatably disposed within a vertical plane which is substantially parallel with the vertical plane occupied by the first post.

The second upstanding post is positioned a sufficient distance away from the first post to permit a user to locate his or her body between the posts, and at a height so that the user grasps and operates the hand crank mounted on the first post, while stooping, thereby positioning predominantly the user's backside toward the second post. The hand crank is operatively connected to the plurality of rotating arms such that the user controls the speed of rotation and the force transferred by the rotating arms. As the user operates the hand crank, the paddles, shoes, or the like mounted on the outboard ends of the rotating arms inflict repetitive blows to the user's buttocks. The amusement apparatus is foldable into a self-containing package for storage or shipping.

BRIEF DESCRIPTION OF SEVERAL VIEWS OF DRAWINGS

Other objects and advantages of the present invention will be recognized from the description of the invention contained herein, including reference to the claims and the drawings in which:

FIG. 1 is a perspective view of one embodiment of an amusement apparatus embodying various features of the present invention;

FIG. 2 is a side view, partial cutaway of the apparatus depicted in FIG. 1 and illustrating a user positioned to operate the apparatus of FIG. 1;

FIG. 3 is a sectional view taken generally along line 3—3 of FIG. 2, illustrating the drive means connecting the crank and drive shaft of the present invention;

FIG. 4 is a sectional view taken generally along line 4—4 of FIG. 2, illustrating the drive means connecting the rotating arms and drive shaft of the present invention;

FIG. 5 is an side view representation of an alternative embodiment of the drive means for connecting the crank and the rotating arms of the present invention;

FIG. 6 is a perspective view of an alternative embodiment of the drive means for connecting the crank and the rotating arms of the present invention;

FIG. 7 is a perspective view of an alternative embodiment of the orientation of the plurality of rotating arms of the present invention; and

FIG. 8 is a perspective view illustrating an alternative drive train of the present invention.

DETAILED DESCRIPTION OF THE INVENTION

In accordance with the present invention, an amusement apparatus is provided that includes a user-operated and controlled apparatus for self-infliction of repetitive blows to the user's buttocks including a plurality of rotating arms bearing flexible extensions for self-paddling the user's buttocks B. FIGS. 1–8, illustrate the amusement apparatus 10 for self-paddling a user U. As illustrated in FIG. 1 and 2, the self-paddling apparatus 10 includes a display platform 12 having a hollow interior and having a first end 14 portion and a second end 16 portion. The platform 12 is constructed of materials adequate to support the weight of at least a few people standing on the platform, such as a user and an observer. In one embodiment, the platform 12 includes at least two generally equally sized subunits that are connect-

Figure 9.7. The specification section of a U.S. utility patent. Accessed September 2009.

the user's buttocks B when the user is positioned with his or her buttocks B facing the plurality of rotating arms of the second post.

A number of associated mechanical components known to those skilled in the art are illustrated in FIGS. 1–8 for completeness. A plurality of mechanically connected belts or chains, gears, cams, and/or rotating shafts known to one skilled in the art are alternatively utilized with the buttocks B paddling apparatus. An alternate embodiment provides a drive train including a continuous belt as illustrated in FIG. 8. The continuous belt **68** encircles first pulley **38** and extends to second pulley **50** along a drive train pathway including a set of third pulleys **70**, **71** proximate the first pulley **38**, and a set of fourth pulleys **72**, **73** proximate the second pulley **50**. The belt **68** descends from the first pulley **38**, proceeds around an outboard third pulley **70**, and extends to an outboard fourth pulley **72**. The belt **68** proceeds around fourth pulley **72**, extends to and partially encircles second pulley **50**, descends along a return pathway around an inboard fourth pulley **73**, extends to an inboard third pulley **71**, and extends to partially encircle first pulley **38**. The continuous belt **68** may be constructed of a high-strength material that withstands wear without significant stretching during operation, such as polyurethane.

One skilled in the art will recognize that a visual display in the form of a LCD display can connect to the drive train and can provide the rotational speed of the rotating arms that strike the user's buttocks B, and/or a display of the number of paddles or shoes contacting the user's buttocks B per unit of time. Mechanical components which support the system are illustrated for clarity only, and other embodiments could be utilized without interfering with the objects and advantages of the present invention.

From the foregoing description, advantages will be recognized by those skilled in the art for the amusement apparatus including a user controlled cranking apparatus positioned on a platform with a drive train delivering rotational movement generated by operation of a crank positioned on a first post, to a plurality of rotating arms bearing flexible extensions for self-paddling the user's buttocks B. One advantage of the amusement apparatus is that the rate of cranking, and the rate of self-paddling is directly controlled by the user, with no significant delays in transmission of rotational movement provided by the user, and no significant delays for resetting the apparatus or repositioning, the user. The amusement apparatus is operated by one user for self-kicking the user's buttocks, or an alternative embodiment allows one user to operate the crank while a second person positions himself to receive a paddling of his buttocks B for entertainment of observers. The amusement apparatus is configured into at least two subunits connected by hinges for ease of assembly and disassembly. An alternate embodiment provides more than two subunits for variable sizing of the platform to fit the space allowed for the apparatus at local fairs, parades, circuses, or other gatherings of persons.

While a preferred embodiment is shown and described, it will be understood that it is not intended to limit the disclosure, but rather it is intended to cover all modifications and alternate methods falling within the spirit and the scope of the invention as defined in the appended claims. One skilled in the art will recognize variations and associated alternative embodiments. The foregoing description should not be limited to the description of the embodiment of the invention contained herein.

6

Having thus described the aforementioned invention, I claim:

1. An amusement apparatus operated and controlled by a user, comprising:

a platform having a first end and a second end;

first post and second posts detachably mounted on said platform, said first post positioned toward said first end and said second post positioned toward said second end of said platform at a distance from said first post sufficient to permit the user to locate therebetween, facing said first post;

said first post having a crank positioned at a height thereon which requires the user to bend forward toward said first post while grasping said crank with both hands, to prominently present his buttocks toward said second post;

said second post including a top end having a plurality of rotating arms detachably mounted thereon, said plurality of rotating arms having a central axis positioned at a height generally level with the user's buttocks;

an outboard end on each of said plurality of rotating arms; and

a drive train operatively interconnecting between said crank and said central axis of said plurality of rotating arms;

whereby as the user bends forward while grasping said crank, the user bends at his waist to predominantly present his buttocks toward said outboard end on each of said plurality of rotating arms, and the user operates said crank to engage said drive train and to rotate said plurality of rotating arms, causing each respective outboard end on each of said plurality of rotating arms to sequentially strike the user's buttocks.

2. The amusement apparatus of claim **1**, wherein each outboard end of each rotating arm includes a pliable paddle mountable thereon.

3. The amusement apparatus of claim **1**, wherein each outboard end of each rotating arm includes a flexible shoe mountable thereon.

4. The amusement apparatus of claim **1**, wherein said drive train comprises:

first pulley means secured to and rotatable by said crank;

second pulley means secured to said plurality of rotating arms;

shaft means extending between said first and second posts and having first and second ends which are disposed proximate said first and second posts, respective;

third pulley means secured to said first end of said shaft;

fourth pulley means secured to said second end of said shaft;

first flexible continuous loop means entrained about said first and third pulleys for simultaneous rotation of said third pulley, said fourth pulley and said shaft upon rotation of said crank and said first pulley connected thereto; and

second flexible continuous loop means entrained about said second and fourth pulleys for rotating said plurality of rotating arms about said central axis upon rotation of said shaft;

whereby the rate of rotation of said plurality of rotating arms is a function of the rate of rotation of said crank to provide user control over both the rate of rotation of said plurality of rotating arms, hence the frequency and degree of force imparted to the user's buttocks upon

Figure 9.8. The claims section of a U.S. utility patent. Accessed September 2009.

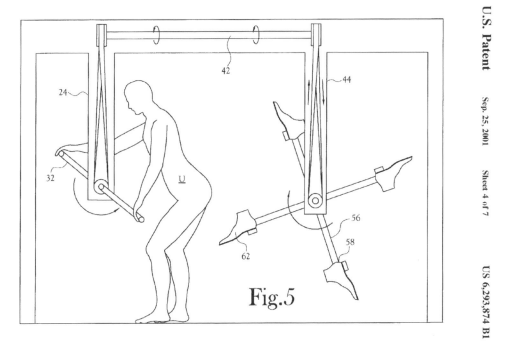

U.S. Patent   Sep. 25, 2001   Sheet 4 of 7   US 6,293,874 B1

Fig.5

Figure 9.9. The drawing section of a U.S. patent. Accessed September 2009.

literature. In sci-tech libraries we refer to searching patents as having the characteristics of walking an English garden maze. The seven steps are:

Classification

1. Brainstorm keywords related to the purpose, use, and composition of the invention.
2. Look up the words in the Index to the U.S. Patent Classification to find potential class/subclasses.
3. Verify the relevancy of the class/subclasses by using the Classification Schedule in the Manual of Classification.
4. Read the Classification Definitions to verify the scope of the subclasses and note "see also" references.

Access Full-Text

5. Search the Issued Patents and the Published Applications databases by "Current US Classification" and access full-text patents and published applications.

Review and References

6. Review the claims, specifications, and drawings of documents retrieved for relevancy.
7. Check all reference and note the "U.S. Cl." and "Field of Search" areas for additional class/subclasses to search.

You will need to allow yourself time to "get lost" and to wear comfortable shoes (or in this case, be in a comfortable chair). The USPTO Web site offers search help in introducing the seven steps; see Figure 9.10.

Once you have reviewed the patent search strategy, your first step is to point your browser to http://www.uspto.gov/. Figure 9.11 displays the homepage of the USPTO.

On the left sidebar of the USPTO main page choose "Patents"; from the pull down list choose "Search Patents." This brings you to the main searching page for locating patent documents. You may search the patent database in a variety of ways depending on the level of information at hand for the search. Search options will include inventor(s) name, the name of a business or institution (known as *assignee*), an issued patent or application number, geographic region, and many more ways. Consider

United States Patent and Trademark Office

Home | Site Index | Search | FAQ | Glossary | Guides | Contacts | eBusiness | eBiz alerts | News | Help

PTDLP > Seven Step Strategy

The 7-Step U. S. Patent Search Strategy

Quick Start

Patent and Trademark Depository Library (PTDL) staff are available to provide training on U.S. patent search processes and research tools including the **Cassis DVD-ROM** system, the **PubWEST** database, and the **USPTO website** .

Classification

1. Brainstorm keywords related to the purpose, use and composition of the invention.

2. Look up the words in the **Index to the U.S. Patent Classification** to find potential class/subclasses.

3. Verify the relevancy of the class/subclasses by using the **Classification Schedule** in the **Manual of Classification** .

4. Read the **Classification Definitions** to verify the scope of the subclasses and note "see also" references.

Access Full-Text

5. Search the **Issued Patents** and the **Published Applications** databases by "Current US Classification" and access full-text patents and published applications.

Review and References

6. Review the claims, specifications and drawings of documents retrieved for relevancy.

7. Check all references and note the "U.S. Cl." and "Field of Search" areas for additional class/subclasses to search.

Figure 9.10. USPTO 7-Step U.S. Patent Search Strategy help page (http://www.uspto.gov/web/offices/ac/ido/ptdl/step7.htm). Accessed September 2009.

Figure 9.11. United States Patent and Trademark Office (USPTO) homepage. Accessed November 2009.

beginning your search with a general topic or keyword. This approach enables the discovery of what has come before in the art related to your area of interest and is a good way to become familiar with the contents of the database and its structure. Using our previous utility patent example, US 6,293,874 B1, *User-Operated Amusement Apparatus for Kicking the User's Buttocks,* we will walk through the seven-step strategy.

Classification System

The classification system of the U.S. Patent and Trademark Office is based on the function of an invention rather than what it may be called, and it is placed into a class relevant to the technology under which it falls. In 1838 the Classification Schedule contained only 21 classes (see Figure 9.12); as of late 2009 there are over 473 classes identified. The patent numbering system is now approaching 75 million (Patent 2009).

Here is one example. I decide to invent a machine to assist in training and improving my tennis game. I design a simple ball-throwing machine that will continuously throw a ball to me so that I may swing my racket at the ball repeatedly as a training technique. I will be calling my new invention a "tennis training tool." What class will my invention fall under? In studying the class that comes up most frequently when conducting my keyword search, I would further investigate Class 473: Games Using Tangible Projectile. No mention of the keyword "tennis" or "training"

US. PATENT OFFICE CLASSIFICATION 1838

Class 1 —AGRICULTURE, INCLUDING INSTRUMENTS AND OPERATIONS

Class 2 —ARTS POLITE, FINE, AND ORNAMENTAL
   Including music, painting, sculpture, engraving, books, paper, printing, binding, jewelry, &c.

Class 3 —CALORIFIC
   Comprising lamps, fire-places, stoves, grates, room-heaters, cooking apparatus, fuel, &c.

Class 4 —CHEMICAL MANUFACTURES, PROCESSES AND COMPOUNDS
   Including medicine, dyeing, color-making, distilling, mortars, cements, &c.

Class 5 —CIVIL ENGINEERING
   Comprising works on rail and common roads, bridges, canals, wharves, docks, rivers, weirs, dams, and other internal improvements

Class 6 —FIBROUS AND TEXTILE SUBSTANCES
   Including machines for preparing and manufacturing the fibres of wool, cotton, silk, fur, &c.

Class 7 —FIRE-ARMS AND IMPLEMENTS OF WAR, AND PARTS THEREOF
   Including manufacture of shot and gunpowder

Class 8 —GRINDING MILLS AND MILL-GEARING
   Containing grain mills, mechanical movements, horse-power, &c.

Class 9 —HYDRAULICS AND PNEUMATICS
   Including water-wheels, wind-mills, and other implements operated by air or water, or employed in the raising and delivery of fluids

Class 10—HOUSEHOLD FURNITURE
   Including domestic implements, washing machines, soap and candle making, bread and cracker machines, feather dressing, &c.

Class 11—LAND CONVEYANCE
   Comprising carriages, cars, and other vehicles, used on roads, and parts thereof

Class 12—LEATHER
   Including tanning and dressing, manufacture of boots, shoes, saddlery, harness, &c.

Class 13—LEVER AND SCREW POWER
   Including presses for packing, expressing balances, windlasses, cranes, jacks, and other mechanical contrivances for raising weight, &c.

Class 14—MATHEMATICAL, PHILOSOPHICAL, AND OPTICAL INSTRUMENTS
   Including clocks, chronometers, &c.

Class 15—MANUFACTURE OF METALS AND INSTRUMENTS THEREFOR
   Including furnaces, implements for casting, nail and screw machines, hardware, safes, cutlery, &c.

Class 16—NAVIGATION AND MARITIME IMPLEMENTS
   Comprising all vessels for conveyance on water, their construction, rigging, and propulsion; implements for fishing; diving-dresses, life-preservers, &c.

Class 17—STEAM AND GAS ENGINES
   Including boilers and furnaces therefor, and parts thereof

Class 18—STONE AND CLAY
   Including stone dressing, clay moulding and burning, mortar machines, &c.

Class 19—SURGICAL INSTRUMENTS
   Including trusses, dental instruments, bathing apparatus, &c.

Class 20—WEARING APPAREL
   Including instruments for manufacturing articles for the toilet, &c.

Class 21—WOOD, MACHINES, AND TOOLS FOR MANUFACTURING
   Including sawing, planing, mortising, shingle, and stave, carpenters' and coopers' implements, buildings, roads, &c.

Class 22—MISCELLANEOUS

Figure 9.12. The Classification Schedule contained 21 classes in 1838.

is retrieved; however, a patent called *Ball Throwing Device* from 1893 is retrieved when the keyword is searched in the database. Performing both a keyword search and a classification search will yield useful results that may be relevant as you conduct your prior art search.

To begin our general search we will need to be on the USPTO Patent Full-Text and Full-Page Image Databases search page (see Figure 9.15). At the bottom of the page you will find a list called "Related USPTO Resources." Link to the Searching by Patent Classification site as in Figure 9.14, USPTO Classification main menu search tools page. You will notice at the top of Figure 9.14 there are menu items listed across the top of the page; choose "Index to the U.S. Patent Classification" and follow the instructions under "Classification Tools."

Classification Tools

- Conceptualize and record keywords to describe the purpose of the invention you wish to search ("buttocks," "self-kicking," "paddle," rotating arms," "repetitive blows," and so on).
- Using the Index to the U.S. Patent Classification (see Figure 9.13: Index to the Patent Classification System), look up your keywords and make a note of the possible class/subclasses that your invention may fall into. You can also perform simple or advanced keyword searches on the USPTO's main search form.
- Check your recorded class/subclass numbers in the Classification Schedule, located in the *Manual of Classification* (472/51, 55, 137 and 482/51, 72, 148) (see Figure 9.14).
- Take time to read the Classification Definitions for the classes that you feel most closely fit your invention description taking care to check all "see also" links. Definitions are located on the same resource page as Figure 9.14.

Acquire Full-Text Patent Documents

From the homepage, click on the "Search Patents" link on the left menu.

- To access and examine the full text of patents that you have determined are of interest, you will need to search the Issued Patents and the Published Applications database by inputting the class/subclass numbers gathered from your search of the Classification Schedule. This is commonly known as "searching prior-art." Figure 9.15 displays the Issued Patent and Patent Application database search engines that are utilized for the retrieval of the full text of patents of interest to your prior art search.

Figure 9.13. Index to the Patent Classification System. Accessed September 2009.

United States Patent and Trademark Office

Home | Site Index | Search | FAQ | Glossary | Guides | Contacts | eBusiness | e

'atents > Guidance, Tools, and Manuals > Classification Main Menu

Class Numbers & Titles | Class Numbers Only | Index to the U.S. Patent Classification (USPC)

A. Access Classification Info by Class/Subclass HELP

1. Enter a US Patent Classification...

[____] **/** [_____]

Class (required)/Subclass (optional)
e.g., 704/1 or 482/1

2. Select what you want...

◉ Class Schedule (HTML)

○ Printable Version of Class Schedule (PDF)

○ Class Definition (HTML)

○ Printable Version of Class Definition (PDF)

○ US-to-IPC Concordance (HTML)

○ US-to-IPC Concordance (PDF)

○ US-to-Locarno Concordance

3. (Submit)   (Reset)

C. Classification Information

- **Information on E-Subclasses**
- **Documents and Reports Related to the Manual of Classification**
- **Classification Orders**
- **USPC Consolidated Glossary** New!

Figure 9.14. USPTO Classification main menu search tools page. Accessed September 2009.

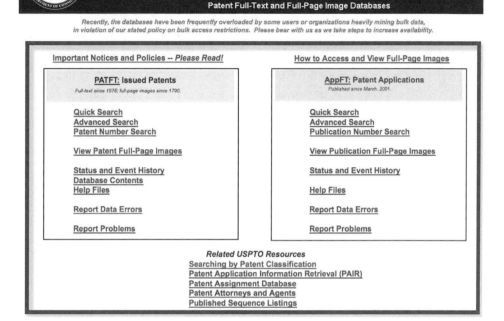

Figure 9.15. USPTO Patent Full-Text and Full-Page Image Databases search page. Accessed November 2009.

Examine Relevant Documents

- Choose from your list of relevant patents and take time to examine the claims, specifications, and drawings for each to determine if they are relevant to your invention.
- Take a close look at the references cited by the inventor as well as taking care to check for additional class/subclass areas noted, as you may need to expand your search beyond the class/subclasses you originally recorded. See Figure 9.16, which displays the subclasses for Class 472. To search for all patents in this class, see Figure 9.17. The search retrieved 73 patents dating back to 1893. And lastly, see Figure 9.18 for one of the first patents classed in 472, Amusement Devices.

Tutorials and videos are freely available on the Internet to assist you in learning how to search the patent literature. Experienced patent literature searchers have created these tools to assist users in learning how to mine the wealth of information buried within. Point your browser to http://www.lib.utexas.edu/engin/patent-tutorial/ tutorial/pattut.html. This is a helpful step-by-step tutorial of the steps necessary to conduct a typical patent search on the USPTO Web site brought to you by the librarians at the University of Texas, McKinney Engineering Library (Baldwin 2008).

A	P	51	**FOR PERPETRATING PRACTICAL JOKE OR INITIATION CEREMONY**
A	P	52	· Unexpectedly expels fluid or powder (e.g., into face of victim, etc.)
A	P	53	·· And produces unexpected noise
A	P	54	· Unexpectedly falls apart or projects movable or free part
A	P	55	·· User contacted by attached projected part
A	P	56	· Produces unexpected noise, vibration, or electric shock

Figure 9.16. Class Schedule for Class 472, Amusement Devices. By clicking the "P" symbol, a search is performed that will retrieve all U.S. patents to be classed in Class 472/Subclass 51. Accessed September 2009.

A helpful agency for patent searchers interested in the search tools of the USPTO is the Patent and Trademark Depository Library Association, established in 1983. According to their Web site, their mission is:

> The Association . . . objectives are to discover the interests, needs, opinions, and goals of the Patent and Trademark Depository Libraries (PTDLs), and to advise the United States Patent and Trademark Office (USPTO) in these matters for the benefit of PTDLs and their users, and to assist the USPTO in planning and implementing appropriate services for the PTDLs, their staffs and patrons.

Along with holding training seminars and patent-related events and conferences, the PTDL hosts a Web site for the deposit of helpful search tools and handouts developed by its members for use by the public.

Additional Patent Search Engines

So far we have focused on the patenting process in the United States and the official patent granting body, the United States Patent and Trademark Office (USPTO), an agency of the U.S. Department of Commerce administered by the Under Secretary of Commerce for Intellectual Property and Director of the USPTO, along with the Commissioner for Patents. There are many international patent-issuing countries that inventors could consider when making application for their invention so as to be granted patent protection more widely than within the United States. This is a common practice, in particular for major institutions and corporations, so it is necessary to expand patent searching knowledge to address global intellectual property searching. This can be achieved by exploring the European Patent Office search Web site (http://www. esp@cenet), a search engine that includes U.S. patents along with 38 European patent-issuing countries. The European Patent Office homepage is pictured in Figure 9.19.

This free database has many nice features. Along with broadening our search globally, it also contains full-text PDF format images of many of the patents, including U.S. patents from 1920 forward (unlike the TIFF format required to view the full-text patents on the USPTO site for 1976 forward). If what you need is a free database that does a good job of searching world patents in one search, you will want to use esp@cenet. Take the Patent Information Tour at http://www.european-patent-office. org/wbt/pi-tour/tour.php.

Searching US Patent Collection...

Results of Search in US Patent Collection db for:
CCL/472/51: 73 patents.
Hits 1 through 50 out of 73

(Final 23 Hits)

(Jump To) [   ]

(Refine Search) | CCL/472/51 |

PAT. NO.   Title
1   7,192,331 **T** Figure kicking toy
2   6,582,314 **T** Attachment for a motor vehicle
3   6,468,126 **T** Pop-up device
4   6,293,874 **T** User-operated amusement apparatus for kicking the user's buttocks
5   6,241,620 **T** Simulated shattered glass novelty device and method of use
6   5,928,085 **T** Novelty gas dispensing nozzle attachment
7   5,842,903 **T** Novelty noise making, odor generating apparatus
8   5,624,320 **T** Flower presentation device
9   5,527,222 **T** Balloon popping device
10 5,399,122 **T** Balloon with accompanying helium supplying cartridge
11 5,389,030 **T** Inflatable novelty device
12 5,385,344 **T** Modular device for playing pranks
13 5,377,380 **T** Simulated vehicle headlight wipers
14 5,188,565 **T** Novelty device and method of using same
15 4,946,745 **T** Novelty statue
16 4,936,582 **T** Golf club

Figure 9.17. Search by Class 472/Subclass 51. Search retrieved 73 patents dating back to 1893. Accessed September 2009.

Google Patents

Google has harnessed the complete holdings of the USPTO patent database (1790 forward) and provides an easy to search and display functionality to that body of literature. No need here to download a TIFF plug-in to capture patent drawings as on the USPTO site; full patents are delivered in easy-to-use PDF format. But searchers beware. This site does not (as of September 2009) include patent applications, international patents, or U.S. patents issued in the previous few months. Google Patents'

Figure 9.18. Patent US 522,900, issued July 10, 1894, from Class 472, Amusement Devices. Accessed September 2009.

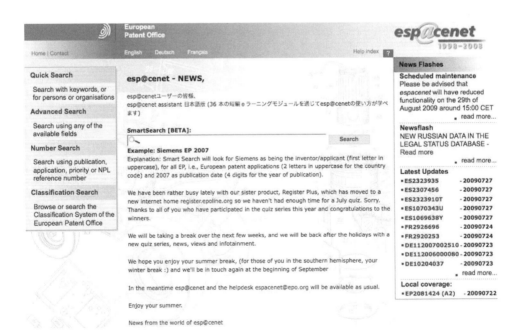

Figure 9.19. European Patent Office search Web site. Accessed September 2009.

Specialized Bibliographies

Government agencies can assist in marketing and advertising new licensable products for the commercial market. One such current awareness tool is the 2009 NASA Patent Abstracts Bibliography (http://www.sti.nasa.gov/Pubs/patents/patents.html). Published annually, it includes citations and abstracts of NASA-owned inventions covered by U.S. patents and patent applications. When available, a key technical illustration is also included.

REFERENCES

Baldwin, Virginia. 2008. "Patent Information in Science, Technical, and Medical Library Instruction." *Science & Technology Libraries* 28 (3): 263–70.

Carlisle, Rodney P. 2004. Scientific American*'s Inventions and Discoveries: All the Milestones in Ingenuity—From the Discovery of Fire to the Invention of the Microwave Oven.* Hoboken, NJ: John Wiley & Sons.

Gordon, Thomas T., and Arthur S. Cookfair. 2000. *Patent Fundamentals for Scientists and Engineers*, 2nd ed. Boca Raton, FL: Lewis Publishers.

Kamil, Idris. 2003. *Intellectual Property: A Power Tool for Economic Growth.* Geneva, Switzerland: World Intellectual Property Organization.

Patent. 2009. *Encyclopaedia Britannica.* Encyclopaedia Britannica Online. Available at: http://search.eb.com/eb/article-9058705.

WEB SITES

Canadian Intellectual Property Office: http://www.cipo.ic.gc.ca/eic/site/cipointernet-interne topic.nsf/eng/Home

Crazy Patents!: http://www.freepatentsonline.com/crazy.html

European Patent Information Tour: http://www.european-patent-office.org/wbt/pi-tour/tour.php

European Patent Office: http://ep.espacenet.com/

The European Patent Office Guide to Intellectual Property Information and Services on the Internet: http://www.epo.org/topics/ip-webguide.html

Free Patents Online: http://www.freepatentsonline.com/

German Patent and Trademark Office: http://www.dpma.de/english/index.html

Google Patents: http://www.google.com/patents

Japan Patent Office: http://www.jpo.go.jp/

Patent and Trademark Depository Library Association: http://www.ptdla.org/

Scitopia: http://www.scitopia.org/scitopia/

United States Patent and Trademark Office (USPTO): http://www.uspto.gov/

Wacky Patent of the Month: http://www.colitz.com/site/wacky_new.html

ADDITIONAL READINGS

Baldwin, Virginia, ed. 2004. *Patent and Trademark Information: Uses and Perspectives.* Binghamton, NY: The Haworth Press.

Brockman, John, ed. 2001. *The Greatest Inventions of the Past 2000 Years.* London: Orion Publishing Group.

Brown, A. E., and H. A. Jeffcott, Jr. 1970. *Beware of Imitations! Absolutely Mad Inventions.* New York: Dover Publications.

Campbell, Hannah. 1964. *Why Did They Name It . . . ?* New York: Fleet Publishing.

Erdmann, Charlotte A. 2002. "Finding Your Way through the Patent Information Maze." *DttP* 30 (3/4): 18–20.

Haven, Kendall. 2005. *100 Greatest Science Inventions of All Time.* Santa Barbara, CA: Libraries Unlimited.

Ikenson, Ben. 2004. *Patents: Ingenious Inventions: How They Work and How They Came to Be.* New York: Black Dog and Leventhal Publishers.

Levy, Joel. 2002. *Really Useful: The Origins of Everyday Things.* Buffalo, NY: Firefly Books.

Morris, Evans. 2004. *From Altoids to Zima: The Surprising Stories Behind 125 Brand Names.* New York: Simon & Schuster.

Mostert, Frederick W., and Lawrence E. Apolzon. 2007. *From Edison to iPod: Protect Your Ideas and Make Money.* New York: DK.

O'Connell, Donal. 2008. *Inside the Patent Factory: The Essential Reference for Effective and Efficient Management of Patent Creation.* Hoboken, NJ: John Wiley & Sons.

Philbin, Tom. 2005. *The 100 Greatest Inventions of all Time.* New York: Citadel.

Rogers, David E., and Amy L. Hartzer. 2008–2009. *Business Success through Innovation: An Insider's Guide to the World of United States Patents.* Scottsdale, AZ: IsoPatent.

Van Dulken, Stephen. 2000. *Inventing the 20th Century: 100 Inventions That Shaped the World: From the Airplane to the Zipper.* Washington Square, NY: New York University Press.

Ward, Randall K. 2005. "From Scientist/Engineer to *Patent* Searcher: Why, What, and How?" *Searcher* 13 (6): 28–31.

Wolff, Thomas E. 2008. "Freedom-to-Operate Patent Searching." *Searcher* 16 (5): 34–39.

Zhang, Li. 2009. "Developing a Systematic Patent Search Training Program." *The Journal of Academic Librarianship* 35 (3): 260–66.

Standards and Specifications

The American Association of Engineering Societies (AAES) provides the following definition of standards:

> Standards provide technical definitions and guidelines for designers and manufacturers. They serve as a common language, defining quality and establishing safety criteria. In the United States, consensus standards are developed by the private sector standards organizations. These standards are used by industry and frequently adopted by government agencies as a means to satisfying regulatory requirements. (AAES, http://www.aaes.org/)

Products, machines, methods, business practices, and telecommunication are devices that you use in your daily life that have been regulated by a standard. A simple way to think about the benefits of setting standards is to consider that it is in our best interest to ensure that things work properly, safely, and efficiently to the benefit of society. By providing specifications and guidelines for machines, structures, and processes, as crafted by professionals and stakeholders in a particular field, universal practices provide such assurances that common practice is regulated in industry. Standards come about as each industry creates working committees to address the need for the creation of a standard for a manufactured item, a process, or a service. These committees are populated by experts in their field of expertise sharing knowledge to create the standard for that product, process, or service.

As an example, the standard for Concrete Radiation Shields, ANSI N101.6-1972, was developed to cover the requirements and recommended practices for the construction of concrete radiation shielding structures and for certain elements of design that relate to problems unique to this type of structure. The information contained in this standard could be applied to a shield for a small item such as a single laboratory workstation or to a large radiochemical plant. The application and regulation of such a standard could save lives and promotes a better standard of living for all of us. As an example of a standard that could apply at home, a common household ladder conforms to a standard and must be manufactured within the specifications of the standard that has been approved, thereby insuring that each household ladder will be protected by the manufacturer's compliance to the standard specification for manufacturing a ladder that follows the guidelines for responsible structural design.

A true story: At precisely 5:27 P.M., November 9, 1965, the electric power for 40,000 square miles of the northeastern United States suddenly went off. In over 10,000 elevators more than a quarter of a million people hung suspended in midair, some as much as a tenth of a mile up. Not one elevator fell. Not a single person was injured or killed. All were saved by a 534-page document called *Safety Code for Elevators and Escalators,* published by the American Society of Mechanical Engineers. Every elevator that is designed, manufactured, installed, or operated in the United States must conform to this standard. Of all those trapped that November day, probably fewer than 25 knew what saved their lives (Batik 1992). Standards affect all areas of our daily lives, and yet the work done by standards agencies is largely unknown to the general public despite the fact that standard practices have been in place since antiquity. Figure 10.1 illustrates an example of how standards affect our everyday lives.

Just about every industrialized nation produces standards, not unlike patents as we discussed in the previous chapter. An important difference to point out is that while patents are the front end of the design and innovation process, standards are the trailing

Figure 10.1. Standardization Essentials: A cartoon by Pascal Krieger, Graphic Artist, ISO Central Secretariat. Reprinted with the permission of the artist and the ISO Central Secretariat.

end, as new products and processes would have already been introduced in the marketplace. The next logical and critical step would be to regulate the manufacture of new inventions and processes by imposing standard safety practices on their design and ultimate use.

Thousands of standards are out there and just as many standard-creating bodies. The umbrella organization in the United States is the American National Standards Institute (ANSI, http://www.ansi.org) (see Figure 10.2). Created in 1928 during the "golden age of standards" and known as the American Standards Association (ASA), one of its first approved standards was the standardization of film speed numbers that were printed on every box of camera film and quickly became familiar to the general public. ANSI oversees and collects standards from issuing bodies. As illustrated by our earlier Concrete Radiation Shield example, this standard was developed by the American Nuclear Society but was facilitated in its formation, approved by, and is available through the ANSI clearinghouse. ANSI aids in the vetting process by acting as the neutral venue for all interested stakeholders and by overseeing and insuring that the process to standardization is a fair one for all parties. Standards are intended as a guide to aid the manufacturer, the consumer, and the general public. Standards are subject to periodic review and users are cautioned to obtain the latest editions. For this 1972 issued standard, a reference librarian would find out if the user required the most recent revision or if they needed a particular issued version for a court case or as a referral to the original document.

Internationally, the organization that performs the same function as ANSI (collecting, approving, and disseminating standards) is the International Organization for Standardization (ISO) (see Figure 10.3). As new inventions are being manufactured

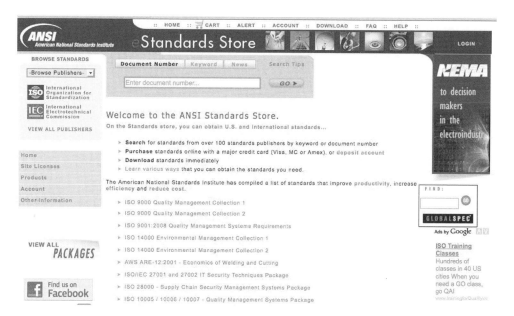

Figure 10.2. American National Standards Institute (ANSI) publications store (http://webstore.ansi.org).

globally (e.g., computers), many international standards are approved that cross country borders. The ISO was formed after World War II; but many countries, including the United States, did not immediately foresee the benefits of the globalization of standards. In fact, many parties were skeptical of foreign-made products and did not support or foster their development. It was not until the approach of the new millennium that worldwide efforts at standardization have taken on urgency and recognition of importance, especially in contributing to increased usability and accessibility of the technologies being developed globally.

Many government agencies have begun to maintain sites that offer free full-text standards documents. One such agency is the U.S. Department of Defense (USDOD). The USDOD is charged with coordinating and supervising all agencies and functions of the government related to national security and the U.S. Armed Forces. As such, an online repository called DTIC, provides information online for the defense community that covers all areas of defense research and includes collections in the

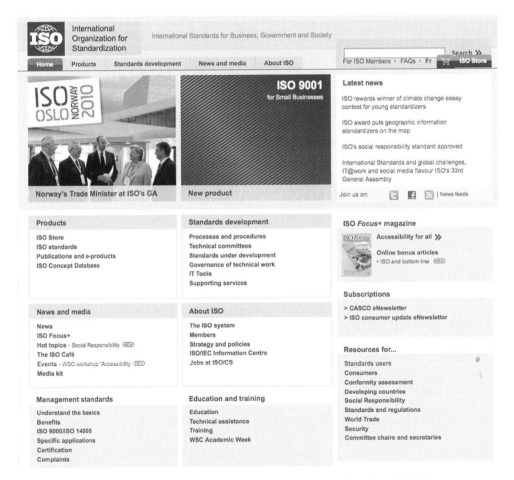

Figure 10.3. International Organization for Standardization (ISO) homepage. Accessed October 2009.

biological and medical sciences, environmental pollution and control, and behavioral and social sciences. Patents, conference and symposia proceedings, and budget information documents are also housed here. DTIC has three levels of access: a public on-line access database called DTIC Online; DTIC Online Access Controlled; and DTIC Online Classified. Entrance to each site is dependent on an individual's registration level. Another collection from DOD available to the general public that includes standards is the DoD Index of Specifications and Standards (http://dodssp.daps.dla.mil/dodiss_index.htm).

Of all of the functions performed by associations, one of the most enduring activities is that of establishing industry standards in their discipline. Standard development is a social process and decisions about acceptable codes of practice come about through discussions published in society journals as well as during conference meetings. By defining what constitutes the professional way of conducting activities and in doing the business of that industry and then making those standards available to the general public, it removes the mystery that so often exists between an industry and the public. What works in a lab must also work in the real world; standards build assurance that this transfer of knowledge is communicated.

MORE ABOUT STAKEHOLDERS

Who are the stakeholders concerned with voluntary and industry standardization? In the U.S. system, nongovernmental standards are voluntary standards. They come about by the meeting of like-minded professionals who are volunteering their time and knowledge to bring about a consensus of acceptable practices to form a standard for others in the same industry to follow. Standards are not mandatory unless they become a part of a regulation. The groups that formulate standards may include companies, professional associations, government agencies, national standards bodies like ANSI, and international standard agencies such as ISO. Of these stakeholders, government agencies are among the major producers and users of standards and specifications.

TYPES OF STANDARDS AND THEIR ELEMENTS

Seven types of standards are identified below and their elements are listed:

- *Dimensional standards* seek to achieve the interchangeability in component parts.

  - Example: SAE J 2440-1998 (SAE J2440-1998): *Domestic Performance Torque Converter Manufacturing.*

- *Materials standards* specify composition, quality, and the chemical and mechanical properties of materials.

  - Example: ASTM E2599-09a: *Standard Practice for Specimen Preparation and Mounting of Reflective Insulation Materials and Radiant Barrier Materials for Building Applications to Assess Surface Burning Characteristics.*

- *Performance standards* set the acceptable levels of efficiency and safety for a product or component.
  - Example: ISO 11620:2008: *Information and documentation— Library performance indicators.*
- *Test methods* recommend tools for comparing the quality or performance of materials and products.
  - Example: ASTM D4309-02(2007): *Standard Practice for Sample Digestion Using Closed Vessel Microwave Heating Technique for the Determination of Total Metals in Water.*
- *Codes of practice* specify procedures for installation, operation, and maintenance for safety and uniform operations.
  - Example: BS M 74:1994: *Air cargo equipment—Basic requirements for aircraft loading equipment.*
- *Terminology and symbols* enable communication across disciplines through the use of standard nomenclature.
  - Example: BS EN 12792:2003: *Ventilation for buildings—Symbols, terminology and graphical symbols.*
- *Documentation standards* spell out the specifications for layout, production, distribution, indexing, and bibliographic description of documents (known to library professionals as "classification standards").
  - Example: ISO 22310:2006: *Information and documentation—Guidelines for standards drafters for stating records management requirements in standards.*

Each standard is organized into sections, and may include some or all of the following category types, depending on the needs of the technical authoring committee:

- Scope
- Significance and use
- Reference documents and standards
- Definitions (terminology)
- Standard specification
- Certification
- Units of measure
- Sample preparation
- Report
- Test methods and procedures
- Comments
- Disclaimer of liability
- Appendix

Standards discuss a range of issues that commonly become the subject of misunderstandings, or in extreme cases contract disputes, and provide guidance for both the

Designation: F 2106 – 02

Reproduced by Global Engineering
Documents With the Permission of
ASTM Under Royalty Agreement

An American National Standard

GLOBAL ENGINEERING DOCUMENTS®

Standard Test Methods for
Evaluating Design and Performance Characteristics of
Motorized Treadmills[1]

This standard is issued under the fixed designation F 2106; the number immediately following the designation indicates the year of original adoption or, in the case of revision, the year of last revision. A number in parentheses indicates the year of last reapproval. A superscript epsilon (ε) indicates an editorial change since the last revision or reapproval.

INTRODUCTION

The goal of these test methods is to provide reliable and repeatable methods for the evaluation of motorized treadmills. Users of the equipment must recognize that conformance to a standard will not necessarily prevent injuries. Like other physical activities, exercise involving a treadmill involves a risk of injury, particularly if the equipment is not maintained or used properly.

1. Scope

1.1 These test methods specify procedures and equipment used for testing and evaluating a motorized treadmill for compliance to Specification F 2115. Both design and operational parameters will be evaluated. Where possible and applicable, accepted test methods from other recognized bodies will be used and referenced. In case of a conflict between this document and Specification F 2115, Specification F 2115 takes precedence.

1.2 *Requirements*—A motorized treadmill is to be tested for all of the following parameters:

1.2.1 Stability,

1.2.2 Exterior design,

1.2.3 Endurance,

1.2.4 Static loading,

1.2.5 Overheating,

1.2.6 Adjustable incline system function,

1.2.7 User interface parameters,

1.2.8 Motorized drive system operation,

1.2.9 Warning label compliance, and

1.2.10 Documentation.

2. Referenced Documents

2.1 *ASTM Standards:*

F 1749 Specification for Fitness Equipment and Facilities Safety Signage and Labels[2]

F 2115 Specification for Motorized Treadmills[2]

2.2 *UL Standards[3]:*

UL 1439 Test for Sharpness of Edges on Equipment

UL 1647 Motor Operated Massage and Exercise Machines

2.3 *European Standard[4]:*

EN 957-1 Stationary Training Equipment—Part 1: General Safety Requirements and Test Methods

2.4 *ISO Standards[5]:*

ISO 4649 Rubber—Determination of Abrasion Resistance Using a Rotating Cylindrical Drum Device

ISO 5904 Gymnastic Equipment—Landing Mats and Surfaces for Floor Exercises—Determination of Resistance to Slipping

3. Terminology

3.1 *Definitions*—For definitions applicable to this standard see Specification F 2115.

4. Significance and Use

4.1 The purpose of these test methods is to provide reliable and repeatable test methods for the evaluation of motorized treadmills assembled and maintained according to the manufacturer's specifications. Use of these test methods in conjunction with Specification F 2115 is intended to insure appropriate performance and reliability of a motorized treadmill and reduce the risk of serious injury from design deficiencies.

5. Certification

5.1 These test methods permit self-certification. It is recommended that each manufacturer employ an independent laboratory to evaluate and validate that their designs and test procedures conform and comply with these test methods and Specification F 2115.

---

[1] This test method is under the jurisdiction of ASTM Committee F08 on Sports Equipment and Facilities and is the direct responsibility of Subcommittee F08.30 on Fitness Products.

Current edition approved May 10, 2002. Published August 2002. Originally published as F 2106 – 01. Last previous edition F 2106 – 01.

[2] *Annual Book of ASTM Standards*, Vol 15.07

[3] Available from Underwriters Laboratories (UL), Corporate Progress, 333 Pfingsten Rd., Northbrook, IL 60062.

[4] Available from CEN Management Centre, 36 rue de Strasse, B-1050, Brussels, Belgium.

[5] Available from International Organization for Standardization (ISO), 1 rue de Varembé, Case postale 56, CH-1211, Geneva 20, Switzerland.

Figure 10.4. A representative front page of a standard from the American Society for Testing and Materials (ASTM). Reprinted, with permission, from ASTM F 2106–02(2010) Standard Test Methods for Evaluating Design and Performance Characteristics of Motorized Treadmills. Copyright ASTM International, 100 Barr Harbor Drive, West Conshohocken, PA.

contractor and the layperson about how such differences can best be avoided. These documents are used in court to settle disputes and to award damages to injured parties. It is critical when assisting users to make sure that you have provided the correct standard for the year in question. Standards are periodically updated and the date of issue will be necessary when determining the practice in place for the period of use when assessing the acceptable voluntary practices for the time period in question. See Figure 10.4 for a representative front page of a standard from the American Society for Testing and Materials (ASTM).

STANDARDS COLLECTIONS

Most major academic and public libraries collect standards issued by national and international bodies. It is an expensive undertaking and one that needs to be consistently updated. Standards are available in a variety of formats and can be purchased based on local needs and usage patterns. The electronic format has become the preferred delivery vehicle of publishers and many are no longer producing print products. If it is not feasible to purchase directly from the producer, another collection option is to purchase standards from a database vendor specializing in providing full-text documents from all standard-issuing bodies. One such vendor is the Information Handling Service, IHS Global (http://global.ihs.com) (see Figure 10.5).

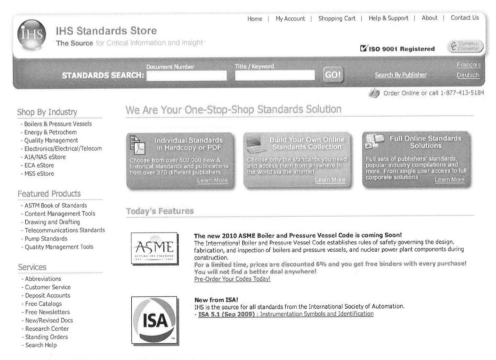

Figure 10.5. Information Handling Service, global standards store (http://www.global.ihs.com). Accessed October 2009.

STANDARDS EDUCATION

For Engineers

Every university engineering program must go through periodic evaluation and accreditation. In the United States this type of accreditation is a nongovernmental peer-review process that educational institutions volunteer to undergo so as to be able to assure students that their program meets the quality standards established by the engineering profession. ABET, Inc., is the agency responsible for the specialized oversight of programs in engineering. For more on the history and structure of ABET, see their Web site (http://www.abet.org).

In the ABET 2007–2008 Criteria for Accrediting Engineering Programs, Proposed Changes document (ABET 2008), it is stated that "students must be prepared for engineering practice through a curriculum culminating in a major design experience based on the knowledge and skills acquired in earlier course work and incorporating appropriate engineering standards and multiple realistic constraints." While this statement is promising, it is generally understood in engineering society circles that standards education is lacking in most U.S. undergraduate programs (Fields 2009). Reasons for this include an already overloaded curriculum that leaves very little time for the implementation of a formal standards education (Schultz 2005). The knowledge of how to apply standards to engineering designs, product development, and business marketing strategies would strengthen the tools a new graduate brings to the job and lessens the time necessary to become familiar with accepted industry practices, proper design protocols, and in product acceptance by the industry. Several university engineering programs have initiated dedicated courses and capstone project work based on standards work as a measure to lessen the gap between educational opportunities and the demands of industry practice.

Additionally, a survey was conducted in 2008 to answer the following question: Do standards education programs have a strategic value? The survey was sent to universities, firms, corporations, standards development organizations, and government departments and agencies. One of the purposes for conducting the study is the estimated rate of engineers' retirements in the near future; it is a prediction that for some nations will mean at least 50 percent of experienced standardization practitioners will no longer be employed. This begs the questions of how will future standardization practitioners be educated and how will past knowledge and expertise be transferred and retained? (Purcell 2008).

For Library Science Professionals

Library professionals also benefit from learning about standards and how they provide the underlying structure for how we organize, share, retrieve, and make sense of our collections. We have standards set out for the proper description of resources in our library collections (cataloging) that we must comply with, such as MARC formats for the bibliographic, authority, holdings, and classification of materials. The Library of Congress, the Library and Archives of Canada, and the British Library serve as the maintenance agency for MARC 21 internationally. They handle all considerations from users of MARC and in their role maintain an electronic bulletin board (MARC@ lov.gov) for broad discussion of proposed changes and other issues for all interested stakeholders around the world. The establishment of an internationally accepted standard for library catalog functionality enables all users on our planet to access resources without borders or limitation.

Consider the power of the International Standard Serial Number (ISSN), ISO 3297-1975; ANSI Z39.3-1979 for the identification of serials. Each serial is assigned a unique publication number, not one that notes the language or country in which it was published, but a number that simply empowers the speedy retrieval of the exact publication as needed. By keeping the system as simplistic as possible it removes the mystery and/or inconvenience of having to navigate unfamiliar systems to locate materials. With the advent of computerized online catalogues, a standard was required to aid in information retrieval among different information systems; thus Z39.50 was born. And in 1977 a standard for preparing bibliographic references (citations, entries in bibliographies, and references in indexes and abstract tools), Z39.29-1977, was implemented. This standard is clear and simple for the scholar to use and works well with MARC formats. Historically this standard has to compete with traditional tools such as *The Chicago Manual of Style,* and for computer scientists, the *IEEE Standards Style Manual*, as well as other publishers' house styles for authors. As professional users of standards ourselves, it is a natural leap to understanding the nature of standards for those we serve in other industry sectors.

For the Public

The American National Standards Institute (ANSI) has a special section of their Web site dedicated to teaching the general public about standards. Called StandardsLearn. org this site offers step-by-step self-paced learning modules for anyone who wishes an introduction to standards and conformity assessment activities. It also provides links to standards activities in the government, legal, and academic worlds.

REFERENCES

ABET Engineering Accreditation Commission. 2008. *Criteria for Accrediting Engineering Programs: Effective for Evaluations during the 2007–2008 Accreditation Cycle.* Baltimore, MD: ABET Engineering Accreditation Commission.

Batik, Albert L. 1992. *The Engineering Standard: A Most Useful Tool.* Ashland, OH: Book Master/El Rancho.

Fields, Rich. 2009. *The Importance of Standards in Engineering.* Presentation. Available at: http://www.ae.msstate.edu/ASC2008/docs/Importance_of_Standards_2008-09-11.ppt.

Purcell, Donald E., ed. 2008. *The Strategic Value of Standards Education.* The Center for Global Standards Analysis. Available at: http://www.ieee.org/documents/Centers_Global_Survey_Report_%28August_2008%29.pdf.

Schultz, Daniel. 2005. "Standards in the Classroom." *ASTM Standardization News,* July. Available at: http://www.astm.org/SNEWS/JULY_2005/schultz_jul05.html.

ADDITIONAL READINGS

American National Standards Institute (ANSI). 2009. Available at: www.StandardsLearn.org.

BSI Library Guide. 2000. *Introduction to Standards.* London: British Standards Institution.

Crawford, Walt. 1986. *Technical Standards: An Introduction for Librarians.* White Plains, NY: Knowledge Industry Publications.

Krechmer, Ken. 2000. "The Fundamental Nature of Standards: Technical Perspective." Standards Report Section. *IEEE Communications Magazine* 38 (6): 70–79.

Mount, Ellis, ed. 1990. *Role of Standards in Sci-Tech Libraries.* New York: Haworth Press.

Paul, J. P. 1997. "Development of Standards for Orthopaedic Implants." *Proceedings of the Institution of Mechanical Engineers* 211, part H.

Ricci, Patricia L. 1992. *Standards: A Resource and Guide for Identification, Selection, and Acquisition.* Woodbury, MN: Pat Ricci Enterprises.

Schultz, Daniel. 2005. "Standards in the Classroom." *ASTM Standardization News,* July. Available at: http://www.astm.org/SNEWS/JULY_2005/schultz_jul05.html.

Spivak, Steven M., and F. Cecil Brenner. 2001. *Standardization Essentials: Principles and Practice.* New York: Marcel Dekker.

Chapter

11

Grey Literature and Technical Reports

GREY LITERATURE

Researchers and scientists strive to be comprehensive in their quest for knowledge on a topic that they are investigating and so must cast their information-seeking net far afield of traditional search resources such as journals and books. This is where grey literature plays the important role of assisting the scholar in gaining a wider perspective on a topic and acts to fill in knowledge gaps (Thompson 2001). Grey literature comes in many formats such as research summaries, topical data sets, special industry and governmental publications, Web sites, newsletters, public industry reports, and patents. In fact, anything other than books and journals may be viewed as grey literature (ACRL STS 2003).

"Grey literature" (GL) has as many definitions as there are forms of publications. The purchase and control of these publications pose problems for information professionals and researchers alike due to a lack of bibliographic control, making them elusive and difficult to locate. The creation and delivery of grey materials can be speedy and dissemination to interested parties is often done via nonprofessional layouts and low print runs, or sent electronically to a small group of identified interested parties. All of these characteristics form road blocks to locating valuable research studies. One definition that is widely used and accepted was created by the Luxembourg Convention on Grey Literature held in 1997 and adopted by the Grey Literature Network Service (1997, 2004), a service that facilitates dialog, research, and communication between persons and organizations in the field of grey literature: "Information produced on all levels of government, academics, business and industry in electronic and print formats not controlled by commercial publishing i.e. where publishing is not the primary activity of the producing body."

The Grey Literature Network Service was founded in 1992 and has set itself the task of identifying and disseminating information on and about grey literature in networked environments. With the advent of the Internet, previously elusive materials such as grey literature are more easily retrieved, thereby making it feasible for more in-depth evaluative scholarly review processes and quality production. It has developed a following of both authors and producers of grey literature and a community of members interested in the formulation of guidelines for the production of scientific and technical reports. Members enjoy benefits such as attendance at the International Conference Series on Grey Literature, the creation and maintenance of Web-based resources, a moderated Listserv, a combined Distribution List, *The Grey Journal* (*TGJ*), as well as curriculum development in the field of grey literature (see Figure 11.1).

A predecessor to the international organization, The Grey Literature Network, a U.S. service, was a portal for technical report information generated through federally funded research and development projects that included five government-sponsored databases. This service was discontinued in October 2007 and the five databases that made up the service were combined and made freely searchable by the U.S. Department of Energy (DOE). It is currently possible to search the combined search engine at: http://Science.gov or http://www.scienceaccelerator.gov/.

Here is a bit of historical perspective: the Internet is now the major finding tool for grey literature but this was not always so. Librarians had to rely on their network of colleagues for locating and obtaining grey materials. They made use of services

Figure 11.1. Grey Literature Network Service Web site. Accessed January 2010.

such as the Linda Hall Library and its predecessor, the Engineering Societies Library collection, the Government Printing Office (GPO), and the National Technical Information Service (for U.S. materials), as well as other interlibrary loan services. To learn about or obtain global documents the information professional relied on national libraries such as the British Lending Library.

Many grey materials were housed in "dark archives" with no published finding tool as a guide. Today, we rely on the open access (OA) movement and the establishment of institutional repositories (IR) to light the way to discovery of grey materials (Bell, Foster, and Gibbons 2005). A survey of grey literature citations using a conventional search tool, the ISI Web of Science, and a nonconventional tool, Google Scholar, validates that Google Scholar is an efficient tool to identify core papers and to track citations from different types of documents. Google Scholar has created a higher visibility for grey literature materials both as cited and citing documents and demonstrates that having a presence on the Web minimizes the difference between conventional and GL materials (Di Cesare, Luzi, and Ruggieri 2008).

Contrary to popular belief, grey literature is subject to a review process. Many types of grey materials such as patents and standards undergo vigorous scrutiny through the process of establishing and passing the requirements of application and acceptance by experts in the discipline(s) they represent. The same is true for technical reports, as these digests from the field are vetted by the academy and the government agencies and corporations that fund them. A survey conducted in 2004 on tracking developments in the field of grey literature notes that 44.2 percent of respondents agreed that grey literature is always subject to a review process (Boekhorst, Farace, and Frantzen 2004).

TECHNICAL REPORTS

Technical reports are considered a form of grey literature that describes the progress or results of scientific or technical research and development. They are usually produced in response to a specific request or research need, and serve as a report of accountability to a funding organization. Such is the typical case today as researchers apply and are granted federal and private funding to study scientific phenomena and to advance new technologies. The practice of recording and reporting activities in the field is today a commonplace and often required task, but this practice has its roots in antiquity.

In ancient times engineers wrote long, detailed accounts of their building projects for their rulers, complete with drawings and procedures for the completion of the work at hand, much like the reports of Frontinus from his lead position as the administrator of the Roman aqueducts at the end of the first century AD. He had discovered faulty intake channels in the original works and is credited with writing a full account, known as *De aquaeductu,* with his thoughts on improvement for the regulation of the water system. Through the careful scribing of early engineers like Frontinus, we witness the birth of technical writing. The prosaically unadorned style of science and technology was the outcome of the technical writing scientific style that arose in the seventeenth century.

Today's technical reports usually fall into three categories: government-sponsored research reports, privately funded research reports, and academia-generated research reports. These will be described in detail after explaining the origins of these reports and how to locate them.

Government agencies, private and public corporations, and academic and research institutes are the main entities producing technical reports. A technical report covers a very narrow topic and is short in length (often under 150 pages). These reports provide graphs and tables as needed, but those in print copies tend not to be of high quality. The ideal method for someone preparing this information is to produce a report quickly to share recent findings with interested researchers much faster than the typical journal or conference proceeding time line would permit. Producers of technical reports have no space limitations, and they often contain very little prose. They are not regulated by a formal-traditional peer review process. Having stated the lack of a traditional peer-reviewed process, it should be noted that peer review of a kind does occur via the very fact that to be approved for a funding contract the original proposal would have been vetted by a committee representing the funding agency. Technical reports may eventually evolve into official journal articles or published as a graduate thesis. Obviously, the length and format would be altered, but the information and ideas would remain relatively the same. It may be the case that you will be asked to obtain a technical report in a particular version/format for a researcher even if the same work has been published elsewhere. It may be crucial for the researcher to see the same work in its various forms for peer review. A careful information professional will be certain when providing copies of technical reports to query as to the particular format/version the user is willing to accept. Both categories may include national or international reports by university departments, institutes, private industry, or government agencies and laboratories.

A technical report is the format of choice for researchers who need to produce a quick scientific update of the progress of their work. Technical reports are generated to share research results with the scientific community and often with the funding agency supporting the work. These reports are the most common form of communication in science and technology; however, there is no clear definition of technical reports as they vary widely in format depending on the discipline. Many of the reports are generated as an obligation for receiving funding from a sponsoring agency; many are produced quarterly as the grant requires.

Technical reports are underutilized in the same way that the patent literature is. They have a great deal of good information; but they are difficult to locate for several reasons. Since they are not always indexed in traditional databases, they are difficult to locate when conducting an online search. In fact, since technical reports do not generally make it into general databases, it can often be the case that they are here one day, and gone the next. Today with the advent of the Internet, many agencies are offering technical reports in full text for free on the Web.

Another reason for the difficulty in locating these is that the publication and dissemination of technical reports have not been centrally coordinated. Therefore, they can be particularly difficult to identify and locate. University technical reports are found mainly by visiting their institution Web site and locating the appropriate department and conducting a search by author. Government reports on the Web can be located by the agency name or through the use of the National Technical Information Service (see Figure 11.2), a fee-based service (such as through a vendor like Cambridge Scientific Abstracts), or by logging on to the free Science.gov: USA.gov for Science portal at http://www.science.gov (see Figure 11.3).

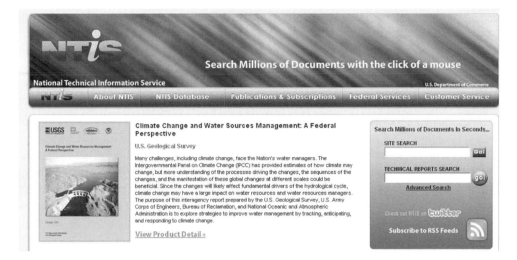

Figure 11.2. National Technical Information Service (NTIS) homepage. Accessed November 2009.

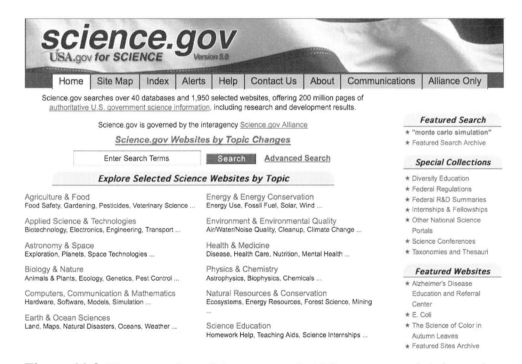

Figure 11.3. Homepage from Science.gov—the U.S. government's information portal. Accessed December 2009.

Technical reports have unique numbering systems depending on the sponsoring agency. NASA is a good example. They have a series of alpha/numeric schemes such as TR for technical reports, or TM for technical memoranda. So a typical number would look like: NASA-TR-50776-1992. The same type of numbering system exists for academic reports as well. To view typical technical reports, see: http://reports-archive.adm.cs.cmu.edu/anon/2003/abstracts/03-102.html.

Given these reasons for difficulty in finding technical reports, how does one locate them? As mentioned previously, many are now available on the Web at the organization's Web site. The database NTIS (National Technical Information Service) is a great place to locate government-funded research reports. The Government Printing Office (GPO) database is also a good place to search. For academic reports, look on the Web at the institution homepage or try in NTIS. Technical reports can also be buried in traditional databases like Engineering Information and INSPEC.

GOVERNMENT-SPONSORED REPORTS

Most reports from agencies that are sponsored by the government may now be found on their Web sites. These agencies and their Web sites are shown below.

National Technical Information Service (NTIS)

NTIS Database covers 1964 to the present. It is the primary resource for accessing U.S. government-sponsored research and worldwide scientific, technical, and engineering information. NTIS is the central source for the sale of unclassified and publicly available information from research reports, journal articles, data files, computer programs, and audiovisual products from federal sources. Information is also available from international government departments and other international organizations including those in Canada, Japan, the former Soviet Union, and both Western and Eastern European countries. The database is accessed via commercial vendors.

Aerospace and High Technology Database

The Aerospace and High Technology database covers 1962 to the present, with about half of the records dated after 1982. As of January 2008, there were 4 million records. The database indexes basic and applied research in aeronautics, astronautics, and space sciences, and also covers technology development and applications in supporting fields such as chemistry, geosciences, physics, communications, and electronics. The main sources are over 3,000 periodicals, conference proceedings, technical reports, trade journal/newsletter items, patents, books, and press releases, but reports by NASA, other U.S. government agencies, international institutions, universities, and private firms may also be included. Access to the database is via CSA Illumina.

PRIVATELY FUNDED REPORTS

Many corporations produce in-house technical reports for proprietary use. These materials may be restricted due to pending patent applications and/or trade secrets.

However, many labs have opened up their collections to academic and public users whenever possible for the greater knowledge of other professionals in the field. Hewlett Packard HP Labs is one such example of a corporation willing to share as feasible to their business model; see: http://www.hpl.hp.com/techreports/.

ACADEMIA-GENERATED RESEARCH REPORTS

Technical reports are also generated as a result of the research within departments of academic institutions. They are one method of reporting conditions in the field (e.g., engineers on a work site) and in the discovery of new software techniques, languages, and hardware component innovations (e.g., computer scientists). In this setting the purpose of technical reports is to create a time stamp of the work and ideas presented so as to protect one's intellectual property in a very competitive environment. It is also a fast medium for communication, not subject to the usual delays with peer review for scholarly publications, and is an informal venue for sharing experimental approaches and hypotheses that may be actively discussed and debated. Academic technical reports are usually available in full text at no cost on the university's Web site or institutional repository. To view an active example, see CMU Technical reports at: http://www.cs.cmu.edu/research/publications/.

Grey literature and technical reports, while elusive and often difficult to obtain, can be a "jewel in the rough" when conducting a comprehensive information retrieval search. Lists of Web sites to explore and become familiar with, from government and non-profits to for-profit agencies, are listed in the references for this chapter. Grey literature is an ever-evolving body of literature that provides access to valuable, nontraditional information sources necessary to advance science and technology globally. This body of work has been elevated in status and is now discoverable due to modern computer technologies.

REFERENCES

Association of College and Research Libraries (ACRL) Science and Technology (STS) Subject and Bibliographic Access Committee. 2003. *Grey Literature: An Annotated Bibliography.* Web document prepared by Bryna Coonin. Available at: URL:http://personal.ecu.edu/cooninb/Greyliterature.htm.

Bell, Suzanne, Nancy Fried Foster, and Susan Gibbons. 2005. "Reference Librarians and the Success of Institutional Repositories." *Reference Services Review* 33 (3): 283–90.

Boekhorst, A. K., D. J. Farace, and J. Frantzen. 2004. "Grey Literature Survey 2004: A Research Project Tracking Developments in the Field of Grey Literature." *GL6 Conference Proceedings: TextRelease.* Available at: http://www.greynet.org/images/GL6,_Page_1.pdf.

Di Cesare, R., D. Luzi, and R. Ruggieri. 2008. "The Impact of Grey Literature in the Web Environment: A Citation Analysis Using Google Scholar." *GL9 Conference Proceedings: TextRelease.* Available at: http://www.greynet.org/images/GL9,_page_49.pdf.

Grey Literature Network Service. 1997. "Perspectives on the Design and Transfer of STI." Presented at GL3: Third International Conference on Grey Literature, Luxembourg, GL '97.

Grey Literature Network Service. 2004. "Work on Grey in Progress." Presented at GL6: Sixth International Conference on Grey Literature, New York. Available at: http://opensigle. inist.fr/handle/10068/697756.

Thompson, Larry A. 2001. "Grey Literature in Engineering." *Science & Technology Libraries* 19 (3/4): 57–73.

ADDITIONAL READINGS

Dijk, E. M. S., C. Baars, A. Hogenaar, and M. van Meel. 2008. "Accessing Grey Literature in an Integrated Environment of Scientific Research Information." *GL9 Conference Proceedings: TextRelease*. Available at: http://www.greynet.org/images/GL9,_page_15.pdf.

Kowalski, Dawn, et al. 2009. *Writing Engineering Technical Reports*: *Writing@CSU, Writing Guides, Civil Engineering Reports*. Colorado State University. Available at: http://writing.colostate.edu/guides/documents/ce-trpt/index.cfm.

Ranger, Sara L. 2005. "Grey Literature in Special Libraries: Access and Use." *Publishing Research Quarterly* 21 (1): 53–63.

Zhu, Qin. 2006. "The Nuts and Bolts of Delivering New Technical Reports via Database Generated RSS Feeds." *Computers in Libraries* 26 (2): 24–28.

WEB SITES

Aerade: Your quality portal to aerospace & defense resources on the Internet: http://aerade.cranfield.ac.uk/reports.html

Aerospace and High Technology Database: http://www.csa.com/factsheets/aerospace-set-c.php

Building and Fire Research Laboratory, Fire Research Information (FIREDOC): http://fire.nist.gov/bfrlpubs/

Congressional Office of Technology Assessment Historical archives: http://www.princeton.edu/~ota/

Contrails: Aerospace History from the Wright Air Development Center Digital Collection: http://contrails.iit.edu/

CRREL Technical Publications, U.S. Army Corps of Engineers: http://www.crrel.usace.army.mil/library/technicalpublications.html

Defense Technical Information Center (DTIC) Online: http://www.dtic.mil/dtic/

GreyNet Literature Network Service: http://www.greynet.org/

Hewlett Packard HP Labs Technical Reports: http://www.hpl.hp.com/techreports/index.html

IBM Research, Technical Paper Search: http://domino.research.ibm.com/library/cyberdig.nsf/index.html

Information Bridge: DOE Scientific and Technical Information: http://www.osti.gov/bridge/

NASA Technical Reports Server (NTRS): http://ntrs.nasa.gov/search.jsp

National Service Center for Environmental Publications (NSCEP): http://www.epa.gov/ncepihom/

National Technical Information Service: http://www.ntis.gov/

Science Accelerator, Office of Scientific and Technical Information: http://www.scienceac celerator.gov/

Science.gov: UA.gov for Science: http://www.science.gov/

Technical Report Archive & Image Library: http://sites.google.com/a/gwla.org/trail/Home

TechXtra: Engineering, Mathematics, and Computing: http://www.techctra.ac.uk/

TRAIL: Technical Report Archive and Image Library: http://digicoll.manoa.hawaii.edu/ techreports/

Virtual Technical Reports Center E-Prints, Preprints, & Technical Reports on the Web: http://www.lib.umd.edu/ENGIN/TechReports/Virtual-TechReports.html

World Resources Institute Publications: http://www.wri.org/publications

World Wide Science: A global science gateway: http://worldwidescience.org/

Biographical Resources

Information on scientists including their educational background, work experience, and accomplishments is a substantial and important component of the scientific literature. Such information may be as simple as a current address and telephone number or other basic facts to longer, more comprehensive entries that provide information on education, current and previous employment, society memberships, research interests, publications including books and journal articles, honors and awards, contributions to science, and personal or family data.

In general, inquiries and requests for biographical information tend to be in one of four categories or groups. They include persons who need information on the famous dead, famous living, obscure dead, and obscure living. Obviously the first two categories are more readily available, but others can be found. An extremely wide variety of publications contain biographical information. Examples include:

- Biographies in Collected Works
- Biographical Information in Dictionaries
- Biographies in Encyclopedias
- Biographical Information in Festschriften
- Biographical Information in Journals and Newspapers
- Information in Membership Directories
- Individual and Collective Biographies of Scientists in Books
- Bibliographies of Biographical Information
- Biographical Databases

Each of these will be expanded below. The category Biographies in Collected Works is further divided by General and Popular, Scholarly, Biographies of Women, and Biographies of African Americans and other Groups.

BIOGRAPHIES IN COLLECTED WORKS

The category of biographies in collected works has numerous titles, and for some titles the information consists basically of a few lines while for other titles individual entries are quite extensive. Some titles cover as few as several hundred individuals while others contain information for more than 100,000 individuals. Some titles deal with all types of scientists and engineers while others are confined to one specific discipline. Other titles in this category are published on a onetime basis while still others are issued as serials frequently published on a regular (annual or biennial) basis or on an irregular schedule.

General and Popular Biographies

1. *Asimov's Biographical Encyclopedia of Science and Technology.* 2nd rev. ed. Isaac Asimov. Garden City, NY: Doubleday, 1982. ISBN: 0385177712.

    The lives and achievements of great scientists from ancient times to the present are chronologically arranged in this title. There are 1,510 biographical entries that are numbered consecutively (e.g., Leeuwenhoek is number 221). Within most entries are various cross-references to other numbered biographies. For example, Malpighi (number 214), Hooke (number 223), Swammerdam (number 224), and Graaf (number 228) are all mentioned in the Leeuwenhoek entry.

2. *Biographical Encyclopedia of Scientists.* Edited by Richard Olson. NY: Marshall Cavendish, 1998. ISBN: 0761470646.

    This five-volume set covers almost 500 representative figures in the history of science from ancient Greeks to contemporary working scientists. Each entry begins with areas of achievement and contribution together with a detailed time line followed by an essay of 750–1,250 words. Entries are usually accompanied by one or more sidebars that detail current research, theories, inventions, etc. A bibliography of books and journal articles by and about the featured individual concludes each entry. Photographs, drawings, and other illustrations are included and complement the text. Entries are signed by the contributing author.

3. *Encyclopedia of World Scientists.* Elizabeth H. Oakes. New York: Facts on File, 2001. ISBN: 081604130X.

    Nearly 500 known scientific greats of history as well as contemporary emerging scientists are included in this title. The collection covers 200 women and minority scientists among the others. Scientists working in fields ranging from astronomy and biology to engineering and mathematics are represented in essays ranging in length from 500 to 1,000 words. There are approximately 100 illustrations. Entries are indexed by field of specialization, country of birth, country of major scientific activity, and year of birth. Also included is a chronology, bibliography, and subject index.

4. *Notable Scientists: From 1900 to the Present.* Edited by Brigham Narins. Farmington Hills, MI: Gale Group, 2001. ISBN: 0787617512.

This five-volume set profiles almost 1,300 scientists who represent the twentieth century's most prolific and productive contributors in all areas of the biological and physical sciences including both pure and applied aspects. This list provides biographies of women scientists; Asian American, African American, Hispanic American, and Native North American scientists; as well as individuals from countries outside North America and Western Europe. Essays range from 750 to 2,500 words and provide personal and career information including selected writings by the scientist as well as suggestions for further information and readings that relate to the particular scientist. There are approximately 400 photographs as well as a general bibliography and time line of scientific achievements. The set has four indexes: field of specialization, gender, nationality/ethnicity, and subject.

5. *The Scientific 100: A Ranking of the Most Influential Scientists, Past and Present.* John Simmons. New York: Citadel, 2000. ISBN: 0806520949.

This volume not only provides information on the 100 most influential scientists from Archimedes to Hawking but ranks all 100 in order of influence. Each entry is four or five pages in length and includes a photograph or sketch, source notes, bibliography, and an index as well as a brief essay entitled "Inexcusable Omissions, Honorable Mentions, and Also-rans." The top 10 scientists in rank order are: (1) Newton, (2) Einstein, (3) Bohr, (4) Darwin, (5) Pasteur, (6) Freud, (7) Galileo, (8) Lavoisier, (9) Kepler, and (10) Copernicus.

Scholarly Biographies

1. *American Men and Women of Science.* 26th ed. Farmington Hills, MI: Gale Cengage, 2009. ISBN: 9781414433004.

This reference set has been published under this title since 1989. It continues *American Men and Women of Science: Physical and Biological Sciences* (1972–1986), which was preceded by *American Men of Science: A Biographical Directory* (1906–1968). This eight-volume set is a comprehensive compendium of biographical profiles of approximately 135,000 living scientists in the United States and Canada. It covers individuals from the physical, biological, and health sciences. Scientists who have made a significant contribution to a particular field of endeavor (conducting high-quality research, publishing in reputable scientific journals, etc.) are eligible for inclusion. Entries provide date of birth, birthplace, field of specialty, education, honorary degrees, current position, professional and career information, awards, memberships, research information, and e-mail, fax, and mailing addresses. The index volume includes an index of disciplines and subspecialties. Scientists are then listed under their states.

2. *Biographical Memoirs: National Academy of Sciences (U.S.).* Washington: National Academy of Sciences, 1877–. ISSN: 0077-2933.

   Begun in 1877, this title is a series of volumes containing the life histories and selected bibliographies of deceased members of the National Academy of Sciences. A portrait is also included. Colleagues familiar with the discipline and the subject's work prepare the essays. The memoirs are a record of the life and work of the most distinguished leaders in the sciences, as witnessed and interpreted by their colleagues and peers. Collectively, these volumes form a biographical history of science in America.

3. *Biographical Memoirs of Fellows of the Royal Society (Great Britain).* London: Royal Society, 1955–. ISSN: 00804606

   This publication, formerly known as *Obituary Notices of Fellows of the Royal Society* and published from 1932 to 1954, is an annual that contains obituaries of Fellows and is published by the Royal Society of London. Prior to 1932, obituaries were published in the *Proceedings of the Royal Society.* The first such published obituaries appeared in 1830. Initially obituaries were read at the anniversary meeting, often by the president himself, and were printed in the record of that meeting. From 1859 onward they appeared in a separate section at the end of the report. Articles were anonymous until the 1880s. The obituaries are often lengthy and detailed and include portraits as well as bibliographies of book and journal publications.

4. *Dictionary of Scientific Biography.* Charles Coulston Gillispie, Editor in Chief. New York: Charles Scribner's Sons, 1970–1980. ISBN: 0684101149.

   This is a scholarly title that was published from 1970 to 1980 in 16 volumes and consists of scholarly biographies of deceased scientists from antiquity to modern times. It is supplemented by the *New Dictionary of Scientific Biography* and an electronic version that includes both publications. The *New Dictionary of Scientific Biography* (ISBN: 9780684313207), edited by Noretta Koertge, was published in eight volumes in 2007 and contains 775 entries. Almost 500 of these entries are completely new and about scientists who died after 1950 and were not included in the original *Dictionary.* Scientists who were alive when the title was first published are excluded. Those individuals who were covered worked in the areas of mathematics, physics, chemistry, biology, and earth sciences. In addition, engineers, technologists, physicians, social scientists, and philosophers can be found there when their work and contributions directly relate to science or mathematics. The biographies are extensive and contain both personal information as well as considerable detail about an individual's scientific contributions. The articles are generally one to five pages but many are much more extensive and all are written by prominent historians of science. The entries list a selection of the original works of the individual as well as a comprehensive list of the secondary literature reflecting older as well as more recent citations. This set was published under the sponsorship of the American Council of Learned Societies with the endorsement of the History of Science Society. In 1981, after the 16-volume set was complete, a 1-volume abridged edition, the *Concise Dictionary of Scientific Biography* (ISBN: 068416650X), was published. A second edition of this title was published in 2001.

Biographies of Women

1. *American Women in Science, 1950 to the Present: A Biographical Dictionary.* Martha J. Bailey. Santa Barbara, CA: ABC-CLIO, 1998. ISBN: 0874369215.

   More than 300 women who have made significant contributions to all areas of science since 1950 are included in this title. Biographical information includes educational background, employment history, honors, and publications, and places the woman's achievements in the appropriate scientific and social contexts with name, profession, and subject indexes. A detailed preface discusses criteria for inclusion and sources consulted to choose names. This title complements an earlier one, also authored by Bailey, entitled *American Women in Science, Colonial Times to 1950: A Biographical Dictionary* (ISBN: 0874367409).

2. *The Biographical Dictionary of Women in Science: Pioneering Lives from Ancient Times to the Mid-20th Century.* Edited by Marilyn Ogilvie and Joy Harvey. New York: Routledge, 2000. ISBN: 0415920388.

   This two-volume set covers approximately 2,500 entries that document both the careers and personal achievements of women scientists in all time periods, all parts of the world, and all subject fields. Each entry contains a biographical essay, a data section of factual information, and a bibliography with selected primary and secondary sources. An extensive introduction discusses purpose, scope, methodology, audience, and use. Scientists are listed by occupation, time period, and country as well as in a subject index.

3. *International Encyclopedia of Women Scientists.* Elizabeth H. Oakes. New York: Facts on File, 2001. ISBN: 0816043817.

   Approximately 500 women scientists from antiquity to the present from all parts of the world are profiled in this title. A general introduction provides a brief overview of the women featured and criteria for inclusion. The standard biographical information is provided for each individual and includes educational background, scientific work and achievements, and awards. Individuals are indexed by field of specialization and nationality. Also included is a list of recommended print and Internet resources on women scientists and a subject index with page references for scientists features in the text.

4. *Notable Women Scientists.* Edited by Pamela Proffitt. Detroit: Gale Group, 1999. ISBN: 0787639001.

   This title contains biographical profiles of almost 500 women from around the world who have made significant contributions to the fields of science from antiquity to the present. Individual essays range from 400 to 2,000 words and frequently include a photograph. In addition to the usual biographical information (such as education, career information, achievements, and awards), a personal statement by the scientist is included whenever possible. In addition to field of specialization, nationality/ethnicity, and subject indexes, there is a time line that includes scientific milestones of the women in this publication as well as significant events in the history of women in science.

Biographies of African Americans and Other Groups

1. *African Americans in Science: An Encyclopedia of People and Progress.* Charles W. Carey Jr. Santa Barbara, CA: ABC-CLIO, 2008. ISBN: 9781851099986.

   This two-volume set provides in-depth information about the scientific contributions of over 100 leading African American scientists. One section includes more than 150 entries on the institutions and organizations dedicated to helping African Americans pursue scientific and technical careers. Examples of these entries are the American Chemical Society, Camille and Henry Dreyfus Foundation, Fisk University, Gates Millennium Scholars (GMS), National Institute of Environmental Health Sciences, Spelman College, and United Negro College Fund. Additional information in the set includes a chronology, discussions of selected scientific fields, and the contributions of black scientists to a particular field (e.g., microbiology). Another section provides issue-centered entries such as the Tuskegee syphilis experiment, an extensive bibliography of both print and online resources, as well as two indexes, a general index, and a selected topical index.

2. *Blacks in Science and Medicine.* Vivian Ovelton Sammons. New York: Hemisphere, 1990. ISBN: 0891166653.

   This was one of the first biographical sources that specifically centered on African American scientists and physicians. It includes 1,500 individuals who have made contributions to various scientific and medical disciplines. Each entry gives birth and death dates, major specialty, education, and organizational affiliations. In many instances, the dissertation title and publications are included. Additional listings of other biographical information are also included. Inventions are listed according to the individual holding the patent. The preface discusses sources consulted and selection criteria. The volume concludes with an extensive bibliography, an index by profession, and a subject index.

3. *Distinguished Asian Americans: A Biographical Dictionary.* Edited by Hyung-chan Kim. Westport, CT: Greenwood, 1999. ISBN: 0313289026.

   While not limited to scientists, many are listed and include Nobel Prize winners. The 166 individuals included may be native or foreign-born Asian. Most entries range from two to three pages and most are accompanied by a photograph. In some instances, personal interviews were conducted to complete the biographies. Appendixes list individuals by fields of professional activity and ethnic subgroups.

4. *Latinos in Science, Math, and Professions.* David E. Newton. New York: Facts on File, 2007. ISBN: 9780816063857.

   This volume profiles more than 175 individuals from the Spanish-speaking world who are involved in the physical and biological sciences, mathematics, and related professions such as medicine and engineering. Each entry contains an introductory paragraph on the individual followed by a profile concentrating on the events of his or her life related to the field of expertise, followed by a list of further readings. The volume concludes with a bibliography and lists of entries by area of activity, year of birth, and ethnicity or country of origin.

5. *Notable Black American Scientists.* Edited by Kristine Krapp. Detroit: Gale, 1999. ISBN: 0787627895.

   Approximately 250 African Americans who have made significant contributions to various sciences, including inventors, researchers, award winners, and educators are discussed in this volume. The entries range from 400 to 2,000 words and are centered on the scientist's life and professional accomplishments. These are followed by a list of books and journal articles by the individual. Suggestions for further reading complete each entry. Photographs accompany many of the biographical entries. There is a time line of scientific milestones of the book's scientists and gender, field of specialization, and subject indexes.

BIOGRAPHICAL INFORMATION IN DICTIONARIES

Biographical information sometimes appears in scientific dictionaries. Even in some specialized subject dictionaries, such as chemistry, brief biographical information can be obtained. In some titles, the information for a particular chemist may consist only of the birth and death dates together with a very brief entry indicating the chemist's contribution to the field. In other titles the information is more extensive. There is detailed information on Josiah Willard Gibbs in *Hawley's Condensed Chemical Dictionary,* including the fact that he is recognized as the father of modern thermodynamics (Lewis 2002).

BIOGRAPHIES IN ENCYCLOPEDIAS

Many encyclopedias in science and technology include biographies. They range from brief sketches to extensive entries frequently accompanied by a photograph as well as a listing of selected publications. In general, living or deceased individuals that are well known at the national and international levels and have made a significant contribution or important discovery in some particular area of science or technology are usually considered for inclusion. In some science encyclopedias, there are separate entries for individual scientists. For example, there are entries for Mayer, Mendel, and Mendeleyev in *Van Nostrand's Scientific Encyclopedia* (Considine 2002). In other titles there are not separate name entries but biographical information is often included in a subject entry. Basic biographical information on Mendeleyev may be found within the subject entry for the periodic table of the elements.

BIOGRAPHICAL INFORMATION IN FESTSCHRIFTEN

Festschriften are collections of essays or articles written in honor of a colleague for some special occasion, such as retirement or a landmark birthday. Most publications of this type usually contain a biographical essay, frequently accompanied by a list of publications that have appeared in peer-reviewed journals and conference proceedings. One specific example of a festschriften is *Changing Perspectives in the History of Science: Essays in Honour of Joseph Needham* (Teich and Young 1973). In addition to 20 contributed articles by prominent scientists, this monograph also contains an extensive bibliography of the books, journal articles, and occasional papers authored by Needham up to 1973.

BIOGRAPHICAL INFORMATION IN
JOURNALS AND NEWSPAPERS

One of the most frequent sources of biographical information is journals and newspapers. Coverage may be of prominent individuals as well as those who are less known.

Less prominent scientists may be covered in the media as a result of having received a special honor or other recognition or may have made an important contribution, discovery, or breakthrough in some field of science or technology, such as genetics or astronomy. In addition to the science coverage in the write-up, basic biographical information including education, current position, and research interests is almost certainly covered. Famous people, in particular, are likely to be included in both popular and scholarly journals and newspapers on various anniversaries of an important contribution. One of the most frequently covered breakthroughs occurred in 1953 with the publication in the April 25 issue of *Nature* on the X-ray crystallographic measurements of the structure of DNA (Watson and Crick 1953). In 2003, this great moment in the history of science and of humanity was celebrated and re-celebrated in popular magazines and newspapers as well as scholarly journals. In a recent special issue of *Smithsonian,* a popular science magazine, 37 individuals under the age of 36 were recognized as innovators in the arts and sciences ("37 under 36" 2007). Specialized journals published by various societies frequently include obituaries of members and well-known scientists and engineers. Newspapers such as *The New York Times* are a good source for interesting and extensive obituaries of prominent individuals in all fields of science, medicine, and technology.

INFORMATION IN MEMBERSHIP DIRECTORIES

Scientific and technical societies maintain membership information that is usually published as an online directory and access to this information is usually limited to members of that society. For many individuals who are members of a particular professional society, this may be the only source of information for a person. Entries usually consist of the briefest of information including name, title, and home or employment address.

INDIVIDUAL AND COLLECTIVE BIOGRAPHIES OF SCIENTISTS IN BOOKS

Science publishing at the popular or nonspecialist level has exploded within the past 25 years. This output includes numerous biographies from contemporary scientists to those who were pioneers in the various fields of science and technology. In the past few years, biographies have covered Roger Bacon (Clegg 2003), Gregor Mendel (Henig 2000), and Francis Crick (Ridley 2006), to name a selected few. Literally, hundreds of scientific biographies have appeared in the past few years alone. Women scientists including Rachel Carson (Lytle 2007) and Rosalind Franklin (Maddox 2002) have been featured. As can be seen from these few examples, book-length biographies are usually restricted to individuals who are nationally or internationally known and very prominent within the scientific and technical community. Collective biographies have also appeared that feature scientists from highly diverse fields (Anton 2000; Chang 2000).

BIBLIOGRAPHIES OF BIOGRAPHICAL INFORMATION

Many biographical publications serve as indexes to biographical information from collections of works that have been analyzed. One well-known publication is an index to almost 7,500 scientists gathered from more than 300 titles (Ireland 1962). Another, more recent title is an index that provides information and biographical references for approximately 2,500 women scientists, physicians, and engineers collected from 130 sources of biographical information (Herzenberg 1986).

Biographical Databases

Numerous biographical sources are available in print format, and many of these same titles are also available in electronic format or a combination of formats. However,

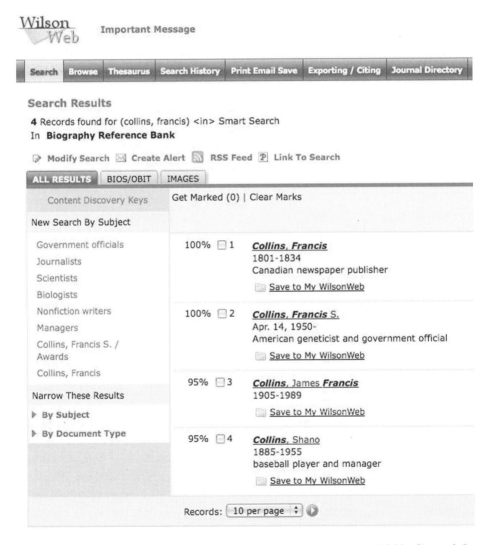

Figure 12.1. Search by named individual. Accessed October 2009. Copyright H. W. Wilson's Biography Reference Bank.

Figure 12.2. Article displayed for biography of Francis Collins. Accessed October 2009. Copyright H. W. Wilson's Biography Reference Bank.

two extensive biographical products are available in electronic format. One of these is the Biography Reference Bank published by H. W. Wilson and another is the Biography Resource Center published by Gale.

The Biography Reference Bank draws from every article about any individual on any of the Wilson Web databases-biographical profiles, feature articles, interviews, essays, book reviews, speeches, or obituaries. It includes *Current Biography, Wilson Biographies Plus Illustrated,* the *World Author* series, *Junior Authors & Illustrators* series, other Wilson biographical works, over 80,000 licensed profiles from major reference publishers, the complete coverage of *Biography Index: 1984 to Present,* plus links to numerous citations and full-text articles (from 3,000 periodicals reviewed and indexed in other Wilson databases). Some 2,000 current books of individuals and collective biography are included each year, as well as biographical materials in otherwise nonbiographical monographs. The database covers more than 550,000 individuals and more than 36,000 images. In addition to these narrative profiles, there are links to almost 400,000 related full-text articles. Scientists, engineers, and related individuals, such as inventors, are included and the narrative profiles are approximately 1,500 words. Searches can be completed by name, profession, place of origin, gender, ethnicity, birth and death dates, keyword, and similar search categories. Another Wilson biographical product is *Biography Index: Past and Present,* which consists of two component databases: *Biography Index Retrospective: 1946–1983* and *Biography Index: 1984 to Present.* Figures 12.1 and 12.2 show two screen shots from H. W. Wilson's Biography Reference Bank. Additional information and free trials are available at: http://www.hwwilson.com/databases.

The Biography Resource Center (Gale) is an online biographical reference database in areas ranging from science and business to the arts and entertainment that contains almost 450,000 biographies on more than 340,000 individuals. It combines more than 135 Gale biographical sources with more than 325 full-text periodicals and journals, upwards of 27,000 portraits, and thousands of recent news briefs on high-interest individuals. This database includes in-depth biographies, brief biographies, full-text magazine articles, Web sites, and additional information for both contemporary and historical individuals. The Biography Resource Center has two add-ons or enhancements for additional biographical information. One is the Marquis Who's

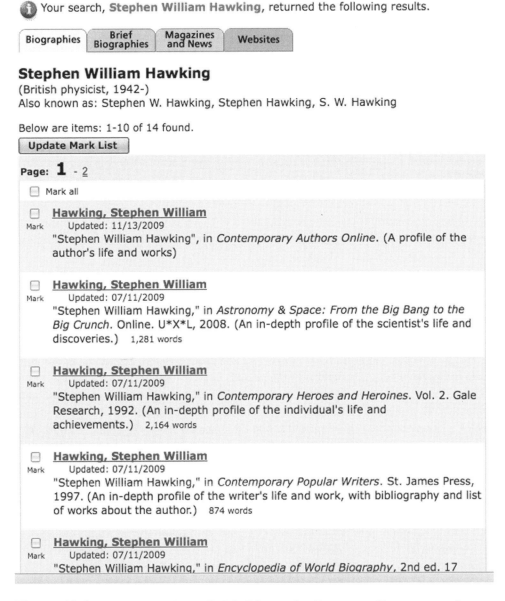

Figure 12.3. Screen shot from Gale's Biography Resource Center; search results on Stephen William Hawking. Accessed October 2009.

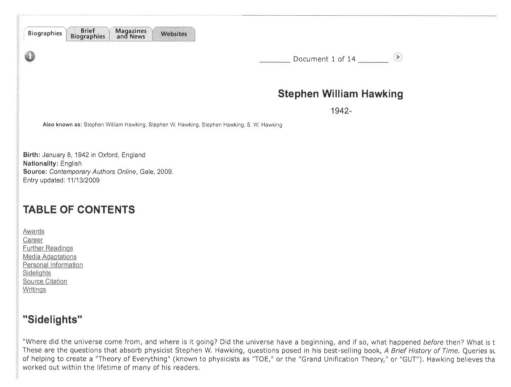

Figure 12.4. Screen shot from Gale's Biography Resource Center of full text of article on Stephen William Hawking. Accessed October 2009.

Who Module that contains more than 1 million brief biographies from all fields of endeavor. A second option is the *Biography Resource Center: Lives & Perspectives Collection,* which contains more than 150,000 biographies drawn from reference titles such as *American Men & Women of Science.* Figures 12.3 and 12.4 show two screen shots from Gale's Biography Resource Center. Additional information and free trials are available at: http://www.Gale.com.

Within the scientific literature, there is a very large quantity of biographical information that covers both living and deceased scientists. Information can be located in a wide variety of sources and the quantity of data, together with the quality of this information, shows considerable variation.

REFERENCES

Anton, Ted. 2000. *Bold Science: Seven Scientists Who Are Changing Our World.* New York: W. H. Freeman.

Chang, Laura. 2000. *Scientists at Work: Profiles of Today's Groundbreaking Scientists from Science Times.* New York: McGraw-Hill.

Clegg, Brian. 2003. *The First Scientist: A Life of Roger Bacon.* New York: Carroll & Graf.

Considine, Glenn D. 2002. *Van Nostrand's Scientific Encyclopedia,* 5th ed. New York: Wiley-Interscience.

Henig, Robin Marantz. 2000. *The Monk in the Garden: The Lost and Found Genius of Gregor Mendel, the Father of Genetics.* Boston: Houghton Mifflin.

Herzenberg, Caroline L. 1986. *Women Scientists from Antiquity to the Present: An Index.* West Cornwall, CT: Locust Hill.

Ireland, Norma Olin. 1962. *Index to Scientists of the World from Ancient to Modern Times: Biographies and Portraits.* Boston: Faxon.

Lewis, Richard J. 2002. *Hawley's Condensed Chemical Dictionary,* 14th ed. New York: John Wiley.

Lytle, Mark Hamilton. 2007. *The Gentle Subversive: Rachel Carson, Silent Spring, and the Rise of the Environmental Movement.* New York: Oxford.

Maddox, Brenda. 2002. *Rosalind Franklin: The Dark Lady of DNA.* New York: Harper Collins.

Ridley, Matt. 2006. *Francis Crick: Discoverer of the Genetic Code.* New York: Harper Collins.

Teich, Mikulas, and Robert Young, eds. 1973. *Changing Perspectives in the History of Science: Essays in Honour of Joseph Needham.* Boston: D. Reidel.

"37 under 36: America's Young Innovators in the Arts and Sciences." 2007. *Smithsonian* (Special Issue) (Fall).

Watson, J. D., and F. H. C. Crick. 1953. "Molecular Structure of Nucleic Acids: A Structure for Deoxyribose Nucleic Acid." *Nature* 171 (4356): 737–38.

Appendix: Subject Bibliographies

The subject bibliographies included in this appendix are selective subject lists of titles grouped in discipline categories. The lists include core titles, seminal works, classics, and standards and represent some of the best resources in a particular subject area. No attempt has been made to be either comprehensive or exhaustive. Formats include guides to the literature, histories, bibliographies, dictionaries, encyclopedias, handbooks and tables, biographies, atlases, electronic databases, and Web sites.

The library science student and the information professional will find the subject guide approach to be useful when building, maintaining, and evaluating subject collections, and the researcher will be assisted in the identification of sources useful to ongoing research activities.

From *Science and Technology Resources: A Guide for Information Professionals and Researchers* by James E. Bobick and G. Lynn Berard. Santa Barbara, CA: Libraries Unlimited. Copyright © 2011.

Anatomy and Physiology

HISTORIES

Persaud, T. V. N. *A History of Anatomy: The Post-Vesalian Era.* Springfield, IL: Charles C. Thomas Publisher, 1997. ISBN: 0398067724.

Rothschuh, Karl E., ed. *History of Physiology.* Huntington, NY: R. E. Krieger Publishing Company, 1973. ISBN: 0882750690.

ELECTRONIC DATABASES

Biosis Previews (1926 to the present). Biosis Previews is a key database for literature searching in biological and life sciences. Covers medical topics with focus on basic research and its coverage of clinical medicine.

ISI Web of Science (1945 to the present, coverage varies). Institute for Scientific Information (ISI) http://scientific.thomson.com/isi/. A multidisciplinary database, with searchable author abstracts, covering the literature of the sciences, social sciences, and the arts and humanities. Users may choose to search across all the databases, or select just one. This unique database indexes and links cited references for each article. All formats provide complete bibliographic data and additional features, such as cited reference searching, links to related articles, plus author and publisher addresses.

Medline (1966 to the present). MEDLINE is the National Library of Medicine's (NLM) premier bibliographic database covering the fields of medicine, nursing, dentistry, veterinary medicine, the health care system, and the preclinical sciences. The MEDLINE file contains bibliographic citations and author abstracts from approximately 3,900 current biomedical journals published in the United States and 70 foreign countries.

ENCYCLOPEDIAS

Blakemore, Colin, and Sheila Jennett, eds. *The Oxford Companion to the Body.* Oxford: Oxford University Press, 2001. ISBN: 019852403X.

Dulbecco, Renato, ed. *Encyclopedia of Human Biology.* San Diego: Academic Press, 1997. 9 vols. ISBN: 0122269705.

Fink, George, ed. *Encyclopedia of Stress.* San Diego: Academic Press, 2000. 3 vols. ISBN: 0122267354.

Knobil, Ernst, and Jimmy D. Neill, eds. *Encyclopedia of Reproduction.* San Diego: Academic Press, 1998. 4 vols. ISBN: 0122270207.

Lerner, K. Lee, and Brenda W. Lerner, eds. *World of Anatomy and Physiology.* Detroit: Thomson/Gale, 2002. 2 vols. ISBN: 0787656844.

Ramachandran, V. S., ed. *Encyclopedia of the Human Brain.* Boston: Academic Press, 2002. 4 vols. ISBN: 0122272102.

ATLASES

Agur, Anne M. R., and Arthur F. Dalley II. *Grant's Atlas of Anatomy,* 11th ed. Philadelphia: Lippincott Williams & Wilkins, 2005. ISBN: 0781742560.

Imwold, Denise, and Janet Parker, es. *Anatomica's Body Atlas.* San Diego: Laurel Glen, 2002. ISBN: 1571459235.

Netter, Frank H. *Atlas of Human Anatomy,* 3rd ed. Teterboro, NJ: Icon Learning Systems, 2003. ISBN: 1929007124.

Olson, Todd R. *PDR Atlas of Anatomy.* Montvale, NJ: Medical Economics Co., 1996. ISBN: 1563632799.

WEB SITES

Arnold's Glossary of Anatomy: http://www.anatomy.usyd.edu.au/glossary/

The InnerBody: Your Guide to the Human Anatomy Online: http://www.innerbody.com/htm/body.html

Master Muscle List: Loyola University Medical Education Network: http://www.meddean.luc.edu/lumen/MedEd/GrossAnatomy/dissector/mml/index.htm

Resources for Teaching Anatomy & Physiology: http://www.csun.edu/science/biology/anatomy/anatomy.html

The Visible Human Project . . . from the National Library of Medicine: http://www.nlm.nih.gov/research/visible/visible_human.html

MISCELLANEOUS

Alexander, R McNeill. *Human Bones: A Scientific and Pictorial Investigation.* New York: Pi, 2005. ISBN: 0131479407.

Brandreth, Gyles. *Your Vital Statistics: The Ultimate Book About the Average Human Being.* Secaucus, NJ: Citadel Press, 1986. ISBN: 0806509805.

Astronomy

GUIDES TO THE LITERATURE

Bruning, David, et al. *Saunders Internet Guide for Astronomy.* Forth Worth, TX: Harcourt Brace College Publishers, 1996. ISBN: 003018858X.

Kemp, D.A. *Astronomy and Astrophysics: A Bibliographical Guide.* Hamden, CT: MacDonald Technical and Scientific, 1970. ISBN: 0208010351.

Lusis, Andy. *Astronomy and Astronautics: An Enthusiast's Guide to Books and Periodicals.* New York: Facts on File, 1986. ISBN: 0816014698.

Seal, Robert A., and Sarah S. Martin. *A Guide to the Literature of Astronomy.* Littleton, CO: Libraries Unlimited, 1977. ISBN: 0872871428.

HISTORIES

DeVorkin, David H. *The History of Modern Astronomy and Astrophysics: A Selected Bibliography.* New York: Garland, 1982. ISBN: 082409283X.

Lang, Kenneth R., and Owen Gingerich. *A Source Book in Astronomy and Astrophysics, 1900–1975.* Cambridge, MA: Harvard University Press, 1979. ISBN: 0674822005.

Lankford, John. *History of Astronomy: An Encyclopedia.* New York: Garland, 1997. ISBN: 081530322X.

From *Science and Technology Resources: A Guide for Information Professionals and Researchers* by James E. Bobick and G. Lynn Berard. Santa Barbara, CA: Libraries Unlimited. Copyright © 2011.

Moore, Patrick. *Patrick Moore's History of Astronomy,* 6th rev. ed. London: Macdonald, 1983. ISBN: 0356086070.

Shapley, Harlow, and Helen E. Howarth. *A Source Book in Astronomy*. New York: McGraw-Hill, 1929. ASIN: B0006AKIVA.

BIBLIOGRAPHIES

Freitag, Ruth S. *Halley's Comet: A Bibliography.* Washington, DC: Library of Congress, 1984. ISBN: 0844404594.

Luther, Paul. *Bibliography of Astronomers: Books and Pamphlets in English by and about Astronomers.* Bernardston, MA: Astronomy Books, 1989. ASIN: B0032D1D4E.

Sunal, Dennis W., and V. Carol Demchik. *Astronomy Education Materials Resource Guide,* 3rd ed. Morgantown, WV: West Virginia University Press, 1985. ISBN: 999760427X.

Tobias, Russell R. *America in Space: An Annotated Bibliography.* Pasadena, CA: Salem Press, 1991. ISBN: 0893566691.

DICTIONARIES

Cambridge Astronomy Dictionary. Cambridge: Cambridge University Press, 1996. ISBN: 0521580072.

Coles, Peter. *The Routledge Critical Dictionary of the New Cosmology.* New York: Routledge, 1999. ISBN: 0415923549. Heck, André. *StarBriefs 2001: A Dictionary of Abbreviations, Acronyms, and Symbols in Astronomy, Related Space Sciences, and Other Related Fields.* Dordrecht: Kluwer, 2001. ISBN: 0792365100.

Illingworth, Valerie. *The Facts on File Dictionary of Astronomy,* 3rd ed. New York: Facts on File, 1994. ISBN: 0816031851.

Ince, Martin. *Dictionary of Astronomy.* Chicago, IL: Fitzroy Dearbon, 1997. ISBN: 1579580181.

Mitton, Jacqueline. *A Concise Dictionary of Astronomy.* New York: Oxford University Press, 1991. ISBN: 0198539673.

Moore, Diane F. *The HarperCollins Dictionary of Astronomy and Space Science.* New York: HarperCollins, 1992. ISBN: 0064610233.

Ridpath, Ian. *A Dictionary of Astronomy.* New York: Oxford University Press, 1996. ISBN: 0521580072.

Schmadel, Lutz D. *Dictionary of Minor Planet Names.* 5th rev. ed. New York: Springer-Verlag, 2003. ISBN: 3540002383.

ELECTRONIC DATABASES

Albrecht, Miguel A., and Daniel Egret. *Databases & On-line Data in Astronomy.* Boston: Kluwer Academic Publishers, 1991. ISBN: 0792312473.

INSPEC (selected earlier literature to 1896, 1898 to the present). Engineering Information, Inc. Contains literature in electrical engineering, electronics, physics, control engineering, information technology, communications, computers, computing, and manufacturing and production engineering. The database contains nearly 10 million bibliographic records taken from 3,850 scientific and technical journals and 2,200 conference proceedings. Approximately 330,000 new records are added to the database annually.

ISI Web of Science (1945 to the present, coverage varies). Institute for Scientific Information (ISI) http://scientific.thomson.com/isi/. A multidisciplinary database, with searchable author abstracts, covering the literature of the sciences, social sciences, and the arts and humanities. Users may choose to search across all the databases, or select just one. This unique database indexes and links cited references for each article. All formats provide complete bibliographic data and additional features, such as cited reference searching, links to related articles, plus author and publisher addresses.

ENCYCLOPEDIAS

Maran, Stephen P. *The Astronomy and Astrophysics Encyclopedia.* Cambridge, UK: Cambridge University Press, 1992. ISBN: 0521417449.

Meyers, Robert A. *Encyclopedia of Astronomy and Astrophysics.* San Diego, CA: Academic Press, 1989. ISBN: 0122266900.

Murdin, Paul. *Encyclopedia of Astronomy and Astrophysics.* Bristol: Institute of Physics, 2001. ISBN: 1561592684.

Parker, Sybil P., and Jay M. Pasachoff. *McGraw-Hill Encyclopedia of Astronomy.* New York: McGraw-Hill, 1993. ISBN: 0070453144.

Shirley, James H., and Rhodes W. Fairbridge. *Encyclopedia of Planetary Sciences.* London: Chapman & Hall, 1997. ISBN: 0412069512.

Weissman, Paul R., et al. *Encyclopedia of the Solar System.* San Diego, CA: Academic Press, 1999. **ASIN:** B0027GY6VY.

HANDBOOKS AND TABLES

Astronomical Almanac. Washington, DC: U.S. Government Printing Office, 2001–Annual. ASIN: B000OHQX1O

Audouze, Jean, and Guy Israel. *Cambridge Atlas of Astronomy,* 3rd ed. New York: Cambridge University Press, 1994. ISBN: 0521434386.

Bakich, Michael E. *The Cambridge Guide to the Constellations.* Cambridge: Cambridge University Press, 1995. ISBN: 0521449219.

Lodders, Katharina, and Bruce Fegley. *The Planetary Scientist's Companion.* New York: Oxford University Press, 1998. ISBN: 0195116941.

Moore, Patrick, and W. I. Tirion. *Cambridge Guide to Stars and Planets.* Cambridge: Cambridge University Press, 1997. ISBN: 0521585821.

Muirden, James. *Amateur Astronomer's Handbook: A Guide to Exploring the Heavens,* 3rd ed. New York: HarperCollins, 1987. ISBN: 0060914262.

Price, Fred W. *The Planet Observer's Handbook,* 2nd ed. Cambridge: Cambridge University Press, 2000. ISBN: 0521789818.

Robinson, Hedley, and James Muirden. *Astronomy Data Book,* 2nd ed. New York: Wiley, 1979. ISBN: 0470265949.ISBN: 0715377434.

Tirion, Wil. *Cambridge Star Atlas,* 2nd ed. New York: Cambridge University Press, 1996. ISBN: 0521560985.

Yearbook of Astronomy. New York: Norton, 1962–Annual. ISSN: 0084-3660.

Zombeck, Martin V. *Handbook of Space Astronomy and Astrophysics,* 2nd ed. Cambridge: Cambridge University Press, 1990. ISBN: 0521345502.

BIOGRAPHY

Abbott, David. *The Biographical Dictionary of Scientists: Astronomers.* New York: P. Bedrick Books, 1984. ISBN: 0584700024.

DIRECTORIES

American Institute of Physics. *Directory of Physics, Astronomy and Geophysics Staff.* College Park, MD: AIP, 1997. ISBN: 1563966654.

American Institute of Physics. *Graduate Programs in Physics, Astronomy and Related Fields.* Melville, NY: AIP, 2011. ISBN: 0735408408.

Heck, André. *StarGuides 2001: A World-Wide Directory of Organizations in Astronomy, Related Space Sciences, and Other Related Fields.* Dordrecht: Kluwer, 2001. ISBN: 0792365097.

Kirby-Smith, H.T. *U.S. Observatories: A Directory and Travel Guide.* New York: Van Nostrand Reinhold, 1976. ISBN: 0442244517.

WEB SITES

AstronomyLINKS: http://astronomylinks.com/

AstroWeb: Astronomy/Astrophysics on the Internet http://www.cv.nrao.edu/fits/www/astronomy.html

National Aeronautics and Space Agency (NASA): http://www.nasa.gov/home/

Science Daily: http://www.sciencedaily.com/news/space_time/astronomy/

Space Telescope Science Institute: http://oposite.stsci.edu/

Biochemistry, Molecular Biology, and Cell Biology

HISTORIES

Echols, Harrison. *Operators and Promoters: The Story of Molecular Biology and Its Creators.* Berkeley: University of California Press, 2001. ISBN: 0520213319.

Harris, Henry. *The Birth of the Cell.* New Haven, CT: Yale University Press, 1999. ISBN: 0300073844.

Judson, Horace F. *The Eighth Day of Creation: Makers of the Revolution in Biology.* Plainview, NY: CSHL Press, 1996. ISBN: 0879694777.

Morange, Michel. *A History of Molecular Biology.* Cambridge, MA: Harvard University Press, 1998. ISBN: 0674398556.

Summers, William C. *Felix d'Herelle and the Origins of Molecular Biology.* New Haven: Yale University Press, 1999. ISBN: 0300071272.

DICTIONARIES

Daintith, John, ed. *The Facts on File Dictionary of Biochemistry.* New York: Facts on File, 2003. ISBN: 0816049149.

Hine, Robert, ed. *The Facts on File Dictionary of Cell and Molecular Biology.* New York: Facts on File, 2003. ISBN: 0816049122.

Lackie, J. M., et al. *The Dictionary of Cell Biology,* 2nd ed. San Diego: Academic Press, 1995. ISBN: 0124325629.

Smith, A. D., ed. *Oxford Dictionary of Biochemistry and Molecular Biology.* Rev. ed. Oxford: Oxford University Press, 2000. ISBN: 0198506732.

Stenesh, J. *Dictionary of Biochemistry and Molecular Biology,* 2nd ed. New York: Wiley, 1989. ISBN: 0471840890.

ELECTRONIC DATABASES

Biosis Previews (1926 to the present). Biosis Previews is a key database for literature searching in biological and life sciences. Covers medical topics with focus on basic research and its coverage of clinical medicine.

Chemical Abstracts (1907 to the present; available electronically via SciFinder Scholar database). SciFinder Scholar is an interface to the *Chemical Abstracts Service (CAS).* It allows you to search millions of chemical references and substances by research topic, author name, structure and substructure, chemical name or CAS Registry Number, company/ name organization, as well as browse the tables of contents in journals, and enable a reaction query.

ISI Web of Science (1945 to the present, coverage varies). Institute for Scientific Information (ISI) http://scientific.thomson.com/isi/. A multidisciplinary database, with searchable author abstracts, covering the literature of the sciences, social sciences, and the arts and humanities. Users may choose to search across all the databases, or select just one. This unique database indexes and links cited references for each article. All formats provide complete bibliographic data and additional features, such as cited reference searching, links to related articles, plus author and publisher addresses.

Medline (1966 to the present). MEDLINE is the National Library of Medicine's (NLM) premier bibliographic database covering the fields of medicine, nursing, dentistry, veterinary medicine, the health care system, and the preclinical sciences. The MEDLINE file contains bibliographic citations and author abstracts from approximately 3,900 current biomedical journals published in the United States and 70 foreign countries.

ENCYCLOPEDIAS

Creighton, Thomas E., ed. *Encyclopedia of Molecular Biology.* New York: Wiley, 1999. 4 vols. ISBN: 0471153028.

Lennarz, William J., and M. Daniel Lane, eds. *Encyclopedia of Biological Chemistry.* Boston: Elsevier Academic Press, 2004. 4 vols. ISBN: 0124437109.

Leung, Albert Y., and Steven Foster. *Encyclopedia of Common Natural Ingredients Used in Food, Drugs, and Cosmetics,* 2nd ed. New York: J. Wiley, 1996. ISBN: 0471508268.

Scott, Thomas, and Mary Eagleson. *Concise Encyclopedia Biochemistry,* 2nd ed. New York: de Gruyter, 1988. ISBN: 0899254578.

WEB SITES

BioChemWeb.org: The Virtual Library of Biochemistry, Molecular Biology and Cell Biology: http://www.biochemweb.org/

Computational Molecular Biology at NIH: http://molbio.info.nih.gov/Index.htm

Geonomics.energy.gov: http://www.ornl.gov/sci/techresources/Human_Genome/posters/chromosome/

National Center for Biotechnology Information: http://www.ncbi.nlm.nih.gov/

Science.gov: http://www.science.gov/

Bioengineering

DICTIONARIES

Bains, William. *Biotechnology from A to Z,* 3rd ed. Oxford; New York: Oxford University Press, 2004. ISBN: 0198524986.

Gosling, Peter J. *Dictionary of Biomedical Sciences.* New York: Taylor & Francis, 2002. ISBN: 0415241383.

ELECTRONIC DATABASES

Biosis Previews (1926 to the present). Biosis Previews is a key database for literature searching in biological and life sciences. Covers medical topics with focus on basic research and its coverage of clinical medicine.

CSA Illustrata: Natural Sciences (1977 to the present). This Cambridge Scientific Abstracts database indexes tables, figures, graphs, charts, and other illustrations from scholarly literature in agriculture, biology, conservation, earth sciences, environmental studies, fish and fisheries, food and food industries, forests and forestry, geography, medical sciences, meteorology, veterinary science, and water resources since 1997.

EI Compendex (also known as the Engineering Index) (1884 to the present). Compendex is the most comprehensive bibliographic database of scientific and technical engineering research available, covering all engineering disciplines. It includes millions of bibliographic citations and abstracts from thousands of engineering journals and conference proceedings. When combined with the Engineering Index Backfile (1884–1969), Compendex covers well over 120 years of core engineering literature.

ISI Web of Science (1945 to the present, coverage varies). Institute for Scientific Information (ISI) http://scientific.thomson.com/isi/. A multidisciplinary database, with searchable author abstracts, covering the literature of the sciences, social sciences, and the arts and humanities. Users may choose to search across all the databases, or select just one. This unique database indexes and links cited references for each article. All formats provide complete bibliographic data and additional features, such as cited reference searching, links to related articles, plus author and publisher addresses.

Knovel: Answers for Science and Engineering (coverage varies). Knovel contains online interactive reference books and databases. It has a database of some of the leading science and engineering reference handbooks, databases, and conference proceedings from publishers such as McGraw-Hill, Elsevier, John Wiley & Sons, ASME, SPE, and ASM International. Copyright: Knovel http://www.knovel.com/

Materials Research Database with Metadex (1966 to the present; Cambridge Scientific Abstracts). This database includes leading materials science databases with specialist content on materials science, metallurgy, ceramics, polymers, and composites used in engineering applications. The collection provides coverage on applied and theoretical materials processes including welding and joining, heat treatment, and thermal spray. Everything from raw materials and refining through processing, welding, and fabrication to end-use, corrosion, performance, and recycling is covered in depth.

Medline (1966 to the present). MEDLINE is the National Library of Medicine's (NLM) premier bibliographic database covering the fields of medicine, nursing, dentistry, veterinary medicine, the health care system, and the preclinical sciences. The MEDLINE file contains bibliographic citations and author abstracts from approximately 3,900 current biomedical journals published in the United States and 70 foreign countries.

ENCYCLOPEDIAS

Akay, Metin. *Wiley Encyclopedia of Biomedical Engineering*. Hoboken, NJ: Wiley-Interscience, 2006. ISBN: 047124967X.

Nalwa, Hari Singh, ed. *Encyclopedia of Nanoscience and Nanotechnology*. Stevenson Ranch, CA: American Scientific Publishers, c2004. ISBN: 1588830012.

Webster, John G. *Encyclopedia of Medical Devices and Instrumentation*. Hoboken, NJ: Wiley-Interscience, 2006. ISBN: 0471263583.

Wnek, Gary E., and Gary L. Bowlin. *Encyclopedia of Biomaterials and Biomedical Engineering,* 2nd ed. Boca Raton, FL: Informa Healthcare, 2008. 4 vols. ISBN: 142007802X.

HANDBOOKS AND TABLES

Bronzino, Joseph D., ed. *The Biomedical Engineering Handbook*. Boca Raton: CRC/Taylor & Francis, 2006. ISBN: 0849321247.

Kutz, Myer. *Standard Handbook of Biomedical Engineering and Design*. New York: McGraw-Hill, 2002. ISBN:0071356371.

2010 Health Devices Sourcebook. ECRI Institute, 2010. ISBN: 0981924123.

Wise, Donald L. *Biomaterials and Bioengineering Handbook.* New York: Marcel Dekker, c2000. ISBN: 0824703189.

BIOGRAPHY

American Men & Women of Science: A Biographical Directory of Today's Leaders in Physical, Biological, and Related Sciences. Detroit, MI: Thomson/Gale, c2003. ISBN: 078766524X.

WEB SITES

American Institute for Medical and Biological Engineering: http://www.aimbe.org/'

The Biomedical Engineering Network: http://www.bmenet.org/BMEnet/

The Biomaterials Network: http://www.biomat.net/index.php?id=1

Bmesource.org: http://171.65.102.151/~bmesource

The Institute of Biological Engineering (IBE): http://www.ibe.org/

MISCELLANEOUS

Ratner, Buddy D., ed. *Biomaterials Science: An Introduction to Materials in Medicine.* Boston: Elsevier Academic Press, 2004. ISBN: 0125824637.

Biotechnology and Genetic Engineering

HISTORIES

Buchholz, Klaus, and John Collins. *Concepts in Biotechnology: History, Science and Business.* New York: Wiley-VCH, 2011. ISBN: 352731766X.

Bud, Robert. *The Uses of Life: A History of Biotechnology.* New York: Cambridge University Press, 1994. ISBN: 0521476992.

Davies, Julian, and William S. Reznikoff. *Milestones in Biotechnology: Classic Papers on Genetic Engineering.* Burlington, MA: Butterworth-Heinemann, 1992. ISBN: 0750692510.

DICTIONARIES

Bains, William. *Biotechnology from A to Z,* 3rd ed. New York: Oxford University Press, 2004. ISBN: 0198524986.

Nill, Kimball. *Glossary of Biotechnology Terms,* 4th ed. Boca Raton, FL: CRC Press, 2005. ISBN: 0849366097.

Steinberg, Mark L., and Sharon D. Cosloy. *The Facts on File Dictionary of Biotechnology and Genetic Engineering,* 3rd ed. New York: Facts on File, 2006. ISBN: 0816063516.

Zhang, Y. H., and M. Zhang. *A Dictionary of Gene Technology Terms.* New York: Informa Healthcare, 2001. ISBN: 185070015X.

ENCYCLOPEDIAS

Flickinger, Michael C., ed. *Encyclopedia of Industrial Biotechnology: Bioprocess, Bioseparation, and Cell Technology,* 2nd ed. New York: Wiley, 2010. ISBN: 0471799300.

Heldman, Dennis R., et al., eds. *Encyclopedia of Biotechnology in Agriculture and Food.* Boca Raton, FL: CRC Press, 2010. ISBN: 0849350271.

Meyers, Robert A., ed. *Molecular Biology and Biotechnology: A Comprehensive Desk Reference.* New York: Wiley-VCH, 1995. ISBN: 047118571X.

Murray, Thomas H., and Max Mehleman. *Encyclopedia of Ethical, Legal, and Policy Issues in Biotechnology.* New York: Wiley-Interscience, 2000. 2 vols. ISBN: 0471176125.

Pandey, Ashok. *Concise Encyclopedia of Bioresource Technology.* Boca Raton, FL: CRC Press, 2004. ISBN: 1560229802.

Post, Stephen Garrard. *Encyclopedia of Bioethics,* 3rd ed. New York: Macmillan Reference USA, 2003. 5 vols. ISBN: 0028657748.

Steinberg, Mark L., and Karen Hubbard. *Encyclopedia of Biotechnology and Genetic Engineering.* New York: Facts on File, 2011. ISBN: 0816067988.

HANDBOOKS AND TABLES

Baltz, Richard H., et al., eds. *Manual of Industrial Microbiology and Biotechnology,* 3rd ed. Washington, DC: ASM Press, 2010. ISBN: 155581512X.

Christou, Paul, and Harry Klee, eds. *Handbook of Plant Biotechnology.* New York: Wiley, 2004. ISBN: 047185199X.

Gad, Shayne Cox, ed. *Handbook of Pharmaceutical Biotechnology.* New York: Wiley-Interscience, 2007. ISBN: 0471213861.

Harisha, S. *Biotechnology Procedures and Experiments Handbook.* Sudbury, MA: Jones & Bartlett Publishers, 2007. ISBN: 1934015113.

Kent, James A. *Kent and Riegel's Handbook of Industrial Chemistry and Biotechnology,* 11th ed. New York: Springer, 2007. ISBN: 0387278427.

Parekh, Sarad R., ed. *The GMO Handbook: Genetically Modified Animals, Microbes, and Plants in Biotechnology.* New York: Humana Press, 2010. ISBN: 1617374822.

MISCELLANEOUS

Fumento, Michael. *BioEvolution: How Biotechnology Is Changing Our World.* San Francisco, CA: Encounter Books, 2003. ISBN: 1893554759.

Grace, Eric S. *Biotechnology Unzipped: Promises and Realities.* Washington, DC: Joseph Henry Press, 2006. ISBN: 0309096219.

Hodge, Russ. *Genetic Engineering: Manipulating the Mechanisms of Life.* New York: Facts On File, 2009. ISBN: 0816066817.

Potter, S. Steven. *Designer Genes: A New Era in the Evolution of Man.* New York: Random House, 2010. ISBN: 140006905X.

Smith, John E. *Biotechnology.* New York: Cambridge University Press, 2004. ISBN: 0521833329.

Stephenson, Frank Harold. *DNA: How the Biotech Revolution Is Changing the Way We Fight Disease.* Amherst, NY: Prometheus Books, 2007. ISBN: 159102482X.

Botany

GUIDE TO THE LITERATURE

Schmidt, Diane, et al. *Guide to Reference and Information Sources in Plant Biology,* 3rd ed. Westport, CT: Libraries Unlimited, 2006. ISBN: 1563089688.

HISTORIES

Blunt, Wilfrid, and William T. Stearn. *The Art of Botanical Illustration: An Illustrated History.* New York: Dover Publications, 1994. ISBN: 0486272656.

Coffey, Timothy. *The History and Folklore of North American Wildflowers.* Boston: Houghton Mifflin, 1994. ISBN: 0395515939.

DICTIONARIES

Bagust, Harold. *The Firefly Dictionary of Plant Names: Common and Botanical.* Buffalo, NY: Firefly Books, 2003. ISBN: 1552976025.

Bailey, Jill, ed. *The Facts on File Dictionary of Botany.* New York: Facts on File, 2003. ISBN: 0816049106.

From *Science and Technology Resources: A Guide for Information Professionals and Researchers* by James E. Bobick and G. Lynn Berard. Santa Barbara, CA: Libraries Unlimited. Copyright © 2011.

Harris, James G., and Melinda Woolf Harris. *Plant Identification Terminology: An Illustrated Glossary,* 2nd ed. Spring Lake, UT: Spring Lake Pub., 2001. ISBN: 0964022176.

Mabberley, David J. *Mabberley's Plant-book: A Portable Dictionary of Plants, Their Classifications, and Uses,* 3rd ed. New York: Cambridge University Press, 2008. ISBN: 0521820715.

DATABASES

CSA Biological Sciences (1982 to the present; Cambridge Scientific Abstracts). Covers resources in life sciences disciplines including cell and molecular biology, neuroscience, biochemistry, entomology, ecology, immunology, microbiology, genetics, immunology, and agriculture.

ISI Web of Science (1945 to the present, coverage varies). Institute for Scientific Information (ISI) http://scientific.thomson.com/isi/. A multidisciplinary database, with searchable author abstracts, covering the literature of the sciences, social sciences, and the arts and humanities. Users may choose to search across all the databases, or select just one. This unique database indexes and links cited references for each article. All formats provide complete bibliographic data and additional features, such as cited reference searching, links to related articles, plus author and publisher addresses.

Medline (1966 to the present). MEDLINE is the National Library of Medicine's (NLM) premier bibliographic database covering the fields of medicine, nursing, dentistry, veterinary medicine, the health care system, and the preclinical sciences. The MEDLINE file contains bibliographic citations and author abstracts from approximately 3,900 current biomedical journals published in the United States and 70 foreign countries.

ENCYCLOPEDIAS

Benvie, Sam. *The Encyclopedia of North American Trees.* Buffalo, NY: Firefly Books, 2000. ISBN: 1552094081.

Couplan, François. *The Encyclopedia of Edible Plants of North America.* New Canaan, CT: Keats Pub., 1998. ISBN: 0879838213.

Marinelli, Janet, ed. *Plant.* New York: DK Pub., 2005. ISBN: 075660589X.

More, David, and John White. *The Illustrated Encyclopedia of Trees.* Portland, OR: Timber Press, 2002. ISBN: 0881925209.

Ness, Bryan D., ed. *Magill's Encyclopedia of Science.* Pasadena, CA: Salem Press, 2003. 4 vols. ISBN: 1587650843.

Robinson, Richard, ed. *Plant Sciences.* New York, Macmillan Reference USA, 2001. 4 vols. ISBN: 002865434X.

HANDBOOKS AND TABLES

Brako, Lois, et al. *Scientific and Common Names of 7,000 Vascular Plants in the United States.* St. Paul, MN: APS Press, 1995. ISBN: 089054171X.

Harris, Marjorie. *Botanica North America: The Illustrated Guide to Our Native Plants, Their Botany, History, and the Way They Shaped Our World.* New York: HarperResource, 2003. ISBN: 0062702319.

Nelson, Lewis, et al. *Handbook of Poisonous and Injurious Plants,* 2nd ed. New York: New York Botanical Garden and Springer, 2007. ISBN: 0387312684.

Phillips, Roger, and Martyn Rix. *The Botanical Garden.* Buffalo, NY: Firefly Books, 2002. 2 vols. ISBN: 1552975916 (vol. 1.); 1552975924 (vol. 2).

Phillips, Roger, et al. *Mushrooms & Other Fungi of North America.* Buffalo, NY: Firefly Books, 2005. ISBN: 1554071151.

Turner, Nancy J., and Patrick von Aderkas. *The North American Guide to Common Poisonous Plants and Mushrooms.* Portland, OR: Timber Press, 2009. ISBN: 9780881929294.

BIOGRAPHY

Isely, Duane. *One Hundred and One Botanists.* Ames: Iowa State University Press, 1994. ISBN: 0813824982.

WEB SITES

Botanical Society of America Online Image Collection: http://images.botany.org/

Flora of North America: http://www.fna.org/

ITIS, the Integrated Taxonomic Information System: http://www.itis.gov/

Plants Database, USDA, Natural Resources Conservation Service: http://plants.usda.gov/index.html

MISCELLANEOUS

Hudler, George W. *Magical Mushrooms, Mischievous Molds.* Princeton, NJ: Princeton University Press, 1998. ISBN: 0691028737.

Raven, Peter H., et al. *Biology of Plants,* 7th ed. New York: W. H. Freeman and Co., 2005. ISBN: 0716710072.

Chemical Engineering

GUIDES TO THE LITERATURE

Hoelscher, Harold Ewald. *Reaction Engineering: A Bibliography and Literature Guide.* New York: American Institute of Chemical Engineers, 1964. ASIN: B0007E3REU.

Yagello, Virginia E., comp. *Guide to Literature on Chemical Engineering.* Washington: American Society for Engineering Education, 1970.

DICTIONARIES

Ashford, Robert D. *Ashford's Dictionary of Industrial Chemicals: Properties, Production, Uses.* London: Wavelength, 1994. ISBN: 0952267403.

Clason, W. E. *Elsevier's Dictionary of Chemical Engineering.* In six languages: English/American, French, Spanish, Italian, Dutch, and German. Amsterdam, New York: Elsevier, 1968. ISBN: 0444407146 (vol. 1); 9780444407146 (vol. 1); 0444407154 (vol. 2); 9780444407153 (vol. 2).

Hackh, Ingo W. D., and Julius Grant. *Hackh's Chemical Dictionary, American and British Usage,* 4th ed. New York: McGraw-Hill, 1969. ISBN: 0070240647.

ELECTRONIC DATABASES

AccessScience. This is the *McGraw-Hill Encyclopedia of Science and Technology* on the Web. It provides full access to articles, dictionary terms, and hundreds of research updates in all areas of science and technology.

Applied Science and Technology Abstracts (ASTA) (1983 to the present; abstracts 1994 to the present). Applied Science and Technology Abstracts covers core English-language scientific and technical publications. Topics include engineering, acoustics, chemistry, computers, metallurgy, physics, plastics, telecommunications, transportation, and waste management. Periodical coverage includes trade and industrial publications, journals issued by professional and technical societies, and specialized subject periodicals, as well as special issues such as buyers' guides, directories, and conference proceedings.

Chemical Abstracts (1907 to the present; available electronically via SciFinder Scholar database). SciFinder Scholar is an interface to the *Chemical Abstracts Service* (*CAS*). It allows you to search millions of chemical references and substances by research topic, author name, structure and substructure, chemical name or CAS Registry Number, company/ name organization, as well as browse the tables of contents in journals, and enable a reaction query.

EI Compendex (also known as the Engineering Index) (1884 to the present). Compendex is the most comprehensive bibliographic database of scientific and technical engineering research available, covering all engineering disciplines. It includes millions of bibliographic citations and abstracts from thousands of engineering journals and conference proceedings. When combined with the Engineering Index Backfile (1884–1969), Compendex covers well over 120 years of core engineering literature.

INSPEC (selected earlier literature to 1896, 1898 to the present). Engineering Information, Inc. Contains literature in electrical engineering, electronics, physics, control engineering, information technology, communications, computers, computing, and manufacturing and production engineering. The database contains nearly 10 million bibliographic records taken from 3,850 scientific and technical journals and 2,200 conference proceedings. Approximately 330,000 new records are added to the database annually.

ISI Web of Science (1945 to the present, coverage varies). Institute for Scientific Information (ISI) http://scientific.thomson.com/isi/. A multidisciplinary database, with searchable author abstracts, covering the literature of the sciences, social sciences, and the arts and humanities. Users may choose to search across all the databases, or select just one. This unique database indexes and links cited references for each article. All formats provide complete bibliographic data and additional features, such as cited reference searching, links to related articles, plus author and publisher addresses.

Knovel: Answers for Science and Engineering (coverage varies). Knovel contains online interactive reference books and databases. It has a database of some of the leading science and engineering reference handbooks, databases, and conference proceedings from publishers such as McGraw-Hill, Elsevier, John Wiley & Sons, ASME, SPE, and ASM International. Copyright: Knovel http://www.knovel.com/

Medline (1966 to the present). MEDLINE is the National Library of Medicine's (NLM) premier bibliographic database covering the fields of medicine, nursing, dentistry, veterinary medicine, the health care system, and the preclinical sciences. The MEDLINE file contains bibliographic citations and author abstracts from approximately 3,900 current biomedical journals published in the United States and 70 foreign countries.

NTIS (National Technical Information Services) Indexes. Cambridge Scientific Abstracts (CSA) Indexes reports on government-sponsored R&D from selected federal agencies (e.g., Department of Energy, EPA), their contractors, and their grantees.

U.S. Patent and Trademark Office Patent Databases (1790 to the present). These databases (http://www.uspto.gov/) allow users to search and view the full-text contents and bibliographic information of patents. Users can search by multiple elements of patent records such as patent numbers, titles, inventors, application dates, descriptions/specifications, and so on. The USPTO houses full text for patents issued from 1976 to the present and TIFF images for all patents from 1790 to the present.

ENCYCLOPEDIAS

Noether, Dorit, and Herman Noether. *Encyclopedic Dictionary of Chemical Technology.* New York: VCH, c1993. ISBN: 0895733293.

O'Neil, Maryadele J. *The Merck Index: An Encyclopedia of Chemicals, Drugs, and Biologicals,* 14th ed. Whitehouse Station, NJ: Merck, 2006. ISBN: 9780911910001

HANDBOOKS AND TABLES

Albright, Lyle Frederick. *Albright's Chemical Engineering Handbook.* Boca Raton, FL: CRC Press, 2009. ISBN: 1420014382.

Chopey, Nicholas P., and Tyler G. Hicks. *Handbook of Chemical Engineering Calculations.* New York: McGraw-Hill, 1984. ISBN: 0070108056.

Himmelblau, David M. *Basic Principles and Calculations in Chemical Engineering.* Upper Saddle River, NJ: Prentice Hall, 1989. ISBN: 013066572X.

Perry, Robert H. *Perry's Chemical Engineers' Handbook.* New York: McGraw-Hill, 2008. ISBN: 9781601196521.

Reynolds, Joseph P. *Handbook of Chemical and Environmental Engineering Calculations.* New York: J. Wiley, 2002. ISBN: 0471402281.

Ullmann, Fritz. *Ullmann's Chemical Engineering and Plant Design.* New York: Wiley-VCH, 2005. ISBN: 9781615833436.

BIOGRAPHIES

Bowden, Mary Ellen. *Chemical Achievers: The Human Face of the Chemical Sciences.* Philadelphia: Chemical Heritage Foundation, 1997. ISBN: 0941901122.

Hall, Paula Quick. *A Day's Work?—A Life's Work! A Visit with Some Women Whose Careers Began with Chemical, Civil, Electrical, Industrial, Mechanical, and Nuclear Engineering.* Washington, DC: Printed and distributed by the Office of Opportunities in Science, American Association for the Advancement of Science, 1987. ASIN: B00072EY6C.

DIRECTORIES

Chem Sources USA, 52nd ed. Clemson, SC: Chemical Sources Intl., 2010. ISBN: 0937020478.

Chemical Week Buyers' Guide. New York: McGraw-Hill. Issued annually as a supplement to *Chemical Week.* ISSN: 0009-272X.

WEB SITES

Chemical Engineer's Resource Page, Online Calculators: http://www.cheresources.com/onlinecalc.shtml

EDA Incorporated, Chemical Engineering: http://www.edasolutions.com/Groups/Chemical Engineering.htm

International Directory of Chemical Engineering: http://www.ciw.unikarlsruhe.de/chem-eng.html

MISCELLANEOUS

Winnick, Jack. *Chemical Engineering Thermodynamics: An Introduction to Thermodynamics for Undergraduate Engineering Students.* New York: Wiley, 1997. ISBN: 0471055905.

Chemistry

GUIDES TO THE LITERATURE

Antony, Arthur. *Guide to Basic Information Sources in Chemistry.* New York: J. Norton Publishers; distributed by Halsted Press, 1979. ISBN: 0470265876.

Bottle, R.T., and J. F. B. Rowland. *Information Sources in Chemistry,* 4th ed. New York: Bowker-Saur, 1993. ISBN: 1857390164.

Maizell, Robert E. *How to Find Chemical Information: A Guide for Practicing Chemists, Educators, and Students,* 3rd ed. New York: John Wiley & Sons, 1998. ISBN: 0471125792.

Mellon, Melvin Guy. *Chemical Publications: Their Nature and Use,* 5th ed. New York: McGraw-Hill, 1982. ISBN: 0070415145.

Skolnik, Herman. *The Literature Matrix of Chemistry.* New York: Wiley, 1982. ISBN: 0471795453.

Wiggins, Gary. *Chemical Information Sources.* New York: McGraw-Hill, 1991. ISBN: 0079099394.

Wolman, Yecheskel. *Chemical Information: A Practical Guide to Utilization,* 2nd rev. and enl. ed. New York: Wiley, 1988. ISBN: 0471917044.

Woodburn, Henry M. *Using Chemical Literature: A Practical Guide.* New York: Marcel Dekker, 1974. ISBN: 0824762606.

From *Science and Technology Resources: A Guide for Information Professionals and Researchers* by James E. Bobick and G. Lynn Berard. Santa Barbara, CA: Libraries Unlimited. Copyright © 2011.

HISTORIES

Asimov, Isaac. *A Short History of Chemistry*. London: Heinemann, 1972. ISBN: 0435550608.

Bensaude-Vincent, Bernadette, and Isabelle Stengers. *A History of Chemistry*. Cambridge, MA: Harvard University Press, 1996. ISBN: 0674396596.

Brock, William Hodson. *The Chemical Tree: A History of Chemistry*. New York: Norton, 2000. ISBN: 0393320685.

Cobb, Cathy, and Harold Goldwhite. *Creations of Fire: Chemistry's Lively History from Alchemy to the Atomic Age*. Cambridge, MA: Perseus Publications, 2001. ISBN: 073820594X.

Ede, Andrew G. *The Chemical Element: A Historical Perspective*. Westport, CT: Greenwood Press, 2006. ISBN: 0313333041.

Greenberg, Arthur. *Chemistry: Decade by Decade*. New York: Facts on File, 2007. ISBN: 0816055319.

Hutton, Keith B. *Chemistry*. Chicago: Fitzroy Dearborn, 2001. ISBN: 1579583598.

Jaffe, Bernard. *Crucibles: The Story of Chemistry from Ancient Alchemy to Nuclear Fission*, 4th ed. New York: Dover Publications, 1976. ISBN: 0486233421.

Meyer, Ernst von. *A History of Chemistry from Earliest Times to the Present Day*. New York: Arno Press, 1975. ISBN: 0405066279.

Morris, Richard. *The Last Sorcerers: The Path from Alchemy to the Periodic Table*. Washington, DC: Joseph Henry Press, 2003. ISBN: 0309089050.

Muir, Matthew Moncrieff Pattison. *A History of Chemical Theories and Laws*. New York: Arno Press, 1975. ISBN: 0405066066.

Multhauf, Robert P. *The Origins of Chemistry*. Langhorne, PA: Gordon and Breach Science Publishers, 1993. ISBN: 2881245943.

Nye, Mary Jo. *Before Big Science: The Pursuit of Modern Chemistry and Physics, 1800–1940*. New York: Twayne Publishers, 1996. ISBN: 080579512X.

Partington, James Riddick. *A History of Chemistry*. Mansfield Center, CT: Martino Pub., 2009. ISBN: 1578987318.

Reese, Kenneth M., ed. *A Century of Chemistry: The Role of Chemists and the American Chemical Society*. Washington, DC: American Chemical Society, 1976. ISBN: 0841203075.

BIBLIOGRAPHIES

Brasted, Robert C., and Leallyn Burr Clapp, eds. *Guidelines and Suggested Title List for Undergraduate Chemistry Libraries*. New York: American Chemical Society, 1982.

Butman, Donna. *Chemistry for Non-Science Trained Librarians and Information Scientists: A Selected Annotated Bibliography*. Silver Spring, MD: CDB Enterprises, 1979. ASIN: B0006DYDDG.

Introducing the Chemical Sciences: A CHF Reading List. Philadelphia, PA: Chemical Heritage Foundation, 1997. ISBN: 0941901181.

Wehefritz, Valentin, ed. *Bibliography on the History of Chemistry and Chemical Technology, 17th to the 19th Century.* München: K. G. Saur, 1994. ISBN: 3598112009.

DICTIONARIES

Bennett, H., ed. *Concise Chemical and Technical Dictionary,* 4th ed. New York: Chemical Publishing Co., 1986. ISBN: 0820602043.

Daintith, John, ed. *The Facts on File Dictionary of Chemistry,* 3rd ed. New York: Facts on File, 1999. ISBN: 0816039097.

Daintith, John, ed. *A Dictionary of Chemistry,* 6th ed. New York: Oxford University Press, 2008. ISBN: 9780199204632.

Hackh, Ingo W. D. *Grant & Hackh's Chemical Dictionary,* 5th rev. ed. by Roger Grant and Claire Grant. New York: McGraw-Hill, 1987. ISBN: 0070240671.

Hampel, Clifford, and Gessner G. Hawley. *Glossary of Chemical Terms,* 2nd ed. New York: Van Nostrand Reinhold, 1982. ISBN: 0442238711.

Lewis, Richard J., Sr., ed. *Hawley's Condensed Chemical Dictionary,* 15th ed. New York: John Wiley & Sons, 2007. ISBN: 9780471768654.

McGraw-Hill Dictionary of Chemistry. New York: McGraw-Hill, 1984. ISBN: 0070454205.

Orchin, Milton, et al. *The Vocabulary and Concepts of Organic Chemistry,* 2nd ed. New York: Wiley-Interscience, 2005. ISBN: 0471680281.

Wertheim, Jane, et al. *The Usborne Illustrated Dictionary of Chemistry.* London: Usborne Pub.; Tulsa, OK: Published in the U.S. by EDC Pub., 2006. ISBN: 0794515606.

ELECTRONIC DATABASES

Bacteriology Abstracts (1982 to the present; ProQuest). Covering topics ranging from bacterial immunology and vaccinations to diseases of humans and animals, the journal provides access to far-reaching clinical findings as well as all aspects of pure bacteriology, biochemistry, and genetics. Main coverage: Bacteriology and microbiology, bacterial immunology, vaccinations, antibiotics, and immunology.

Biosis Previews (1926 to the present). Biosis Previews is a key database for literature searching in biological and life sciences. Covers medical topics with focus on basic research and its coverage of clinical medicine.

Chemical Abstracts (1907 to the present; available electronically via SciFinder Scholar database). SciFinder Scholar is an interface to the *Chemical Abstracts Service (CAS).* It allows you to search millions of chemical references and substances by research topic, author name, structure and substructure, chemical name or CAS Registry Number, company/name organization, as well as browse the tables of contents in journals, and enable a reaction query.

ISI Web of Science (1945 to the present, coverage varies). Institute for Scientific Information (ISI) http://scientific.thomson.com/isi/. A multidisciplinary database, with searchable author abstracts, covering the literature of the sciences, social sciences, and the arts and humanities. Users may choose to search across all the databases, or select just one. This unique database indexes and links cited references for each article. All formats provide complete bibliographic data and additional features, such as cited reference searching, links to related articles, plus author and publisher addresses.

Knovel: Answers for Science and Engineering (coverage varies). Knovel contains online interactive reference books and databases. It has a database of some of the leading science and engineering reference handbooks, databases, and conference proceedings from publishers such as McGraw-Hill, Elsevier, John Wiley & Sons, ASME, SPE, and ASM International. Copyright: Knovel http://www.knovel.com/

Medline (1966 to the present). MEDLINE is the National Library of Medicine's (NLM) premier bibliographic database covering the fields of medicine, nursing, dentistry, veterinary medicine, the health care system, and the preclinical sciences. The MEDLINE file contains bibliographic citations and author abstracts from approximately 3,900 current biomedical journals published in the United States and 70 foreign countries.

Scitopia. Scitopia.org searches more than three and a half million documents, including peer-reviewed journal content and technical conference papers from leading voices in major science and technology disciplines. Searches bibliographic records in each partner's electronic library; patents from the U.S. Patent and Trademark Office, European Patent Office, and Japan Patent Office; and U.S. government documents on the Department of Energy Information Bridge site.

ENCYCLOPEDIAS

Considine, Glenn D., ed. *Van Nostrand's Encyclopedia of Chemistry,* 5th ed. Hoboken, NJ: Wiley-Interscience, 2005. ISBN: 0471615250.

Eagleson, Mary. *Concise Encyclopedia Chemistry.* New York: Walter de Gruyter, 1994. ISBN: 0899254578.

King, Bruce A., ed. *Encyclopedia of Inorganic Chemistry,* 2nd ed. Hoboken, NJ: Wiley, 1994. ISBN: 0470860782.

Kroschwitz, Jacqueline I., ed. *Kirk-Othmer Encyclopedia of Chemical Technology,* 5th ed. Hoboken, NJ: Wiley-Interscience, 2004–2007. 27 vols. ISBN: 0471484946.

Lagowski, J.J., ed. *Chemistry: Foundations and Applications.* New York: Macmillan Reference USA, 2004. 4 vols. ISBN: 0028657217.

McGraw-Hill Concise Encyclopedia of Chemistry. New York: McGraw-Hill, 2004. ISBN: 0071439536.

Rittner, Don, and Ronald A. Bailey. *Encyclopedia of Chemistry.* New York: Facts on File, 2005. ISBN: 0816048940.

Young, Robyn V., ed. *World of Chemistry.* Detroit: Gale Group, 2000. ISBN: 0787636509.

HANDBOOKS AND TABLES

Coyne, Gary. S. *The Laboratory Companion: A Practical Guide to Materials, Equipment, and Technique.* Hoboken, NJ: Wiley-Interscience, 2006. ISBN: 0471780863.

Furr, A. Keith. *CRC Handbook of Laboratory Safety,* 5th ed. Boca Raton, FL: CRC Press, 2000. ISBN: 0849325234.

Lide, David R., er. *CRC Handbook of Chemistry and Physics,* 90th ed. Boca Raton, FL: CRC Press, 2009. ISBN: 9781420090840.

O'Neil, Maryadele J., ed. *The Merck Index: An Encyclopedia of Chemicals, Drugs, and Biologicals,* 14th ed. New York: John Wiley and Sons, 2006. ISBN: 9780911910001.

Speight, James, ed. *Lange's Handbook of Chemistry,* 16th ed. New York: McGraw-Hill, 2005. ISBN: 0071432205.

BIOGRAPHIES

Abbott, David. *The Biographical Dictionary of Scientists, Chemists.* London: Blond Educational, 1983. ISBN: 0584700008.

McGrayne, Sharon Bertsch. *Prometheans in the Lab: Chemistry and the Making of the Modern World.* New York: McGraw-Hill, 2001. ISBN: 0071350071.

Oakes, Elizabeth H. *A to Z of Chemists: Notable Scientists.* New York: Facts on File, 2002. ISBN: 0816045798.

Rogers, David. *Nobel Laureate Contributions to 20th Century Chemistry.* Cambridge: Royal Society of Chemistry, 2006. ISBN: 085404356X.

DIRECTORY

Directory of Graduate Research. Washington, DC: American Chemical Society, Committee on Professional Training, Annual. ISBN: 0841238952.

WEB SITES

ChemTube3D: http://www.chemtube3d.com/

PubChem: http://pubchem.ncbi.nlm.nih.gov/

Science.gov: http://www.science.gov/

The Sheffield Chemdex: The Directory of Chemistry on the WWW since 1993: http://chemdex.org/

Spectral Database for Organic Compounds, SDBS: http://riodb01.ibase.aist.go.jp/sdbs/cgi-bin/cre_index.cgi?lang=eng

MISCELLANEOUS

Ball, Philip. *Stories of the Invisible: A Guided Tour of Molecules.* New York: Oxford University Press, 2001. ISBN: 0192802143.

Cobb, Cathy, and Monty L. Fetterolf. *The Joy of Chemistry: The Amazing Science of Familiar Things.* Amherst, NY: Prometheus Books, 2010. ISBN: 1591027713.

Diagram Group. *Chemistry: An Illustrated Guide to Science.* New York: Chelsea House, 2006. ISBN: 0816061637.

Gray, Theodore. *The Elements: A Visual Exploration of Every Known Atom in the Universe.* New York: Black Dog & Leventhal Publishers, 2009. ISBN: 1579128149.

Heilbronner, Edgar, and Foil A. Miller. *A Philatelic Ramble through Chemistry.* New York: Wiley-VCH, 1998. ISBN: 3906390179.

Karukstis, Kerry K., and Gerald R. Van Hecke. *Chemistry Connections: The Chemical Basis of Everyday Phenomena*, 2nd ed. San Diego, CA: Academic, 2003. ISBN: 0124001513.

Kean, Sam. *The Disappearing Spoon and Other True Tales of Madness, Love, and the History of the World from the Periodic Table of the Elements.* New York: Little, Brown and Co., 2010. ISBN: 0316051640.

Quadbeck-Seeger, Hans Jürgen, ed. *World Records in Chemistry.* New York: Wiley-VCH, 1999. ISBN: 3527295747.

Civil Engineering

HISTORIES

Bruno, Leonard C. *On the Move: A Chronology of Advances in Transportation.* Detroit: Gale Research, 1993. ISBN: 0810383969.

Jackson, Donald C. *Great American Bridges and Dams.* New York: Wiley, 1995. ISBN: 0471143855.

DICTIONARY

Traister, John E. *Illustrated Dictionary for Building Construction.* Lilburn, GA: Fairmont Press, 1993. ISBN: 0881731730.

DATABASES

AccessScience. This is the *McGraw-Hill Encyclopedia of Science and Technology* on the Web. It provides full access to articles, dictionary terms, and hundreds of research updates in all areas of science and technology.

BuildingGreen Suite. The BuildingGreen Suite brings together articles, reviews, and news from *Environmental Building News* (*EBN*) since 1992, product listings from the GreenSpec products directory, and project case studies from the high-performance buildings database.

EI Compendex (also known as the Engineering Index) (1884 to the present). Compendex is the most comprehensive bibliographic database of scientific and technical engineering

From *Science and Technology Resources: A Guide for Information Professionals and Researchers* by James E. Bobick and G. Lynn Berard. Santa Barbara, CA: Libraries Unlimited. Copyright © 2011.

research available, covering all engineering disciplines. It includes millions of bibliographic citations and abstracts from thousands of engineering journals and conference proceedings. When combined with the Engineering Index Backfile (1884–1969), Compendex covers well over 120 years of core engineering literature.

Environmental Sciences and Pollution Management (1967 to the present). This multidisciplinary database, provides unparalleled and comprehensive coverage of the environmental sciences. Abstracts and citations are drawn from over 10,000 serials including scientific journals, conference proceedings, reports, monographs, books, and government publications.

GeoBase (1980 to the present; Elsevier, Inc.). GeoBase is a multidisciplinary database covering human and physical geography, geology, oceanography, geomechanics, alternative energy sources, pollution, waste management, and nature conservation.

GeoRef (1966 to the present). Established by the American Geological Institute in 1966, it provides access to the geoscience literature of the world. GeoRef is the most comprehensive database in the geosciences.

ISI Web of Science (1945 to the present, coverage varies). Institute for Scientific Information (ISI) http://scientific.thomson.com/isi/. A multidisciplinary database, with searchable author abstracts, covering the literature of the sciences, social sciences, and the arts and humanities. Users may choose to search across all the databases, or select just one. This unique database indexes and links cited references for each article. All formats provide complete bibliographic data and additional features, such as cited reference searching, links to related articles, plus author and publisher addresses.

Knovel: Answers for Science and Engineering (coverage varies). Knovel contains online interactive reference books and databases. It has a database of some of the leading science and engineering reference handbooks, databases, and conference proceedings from publishers such as McGraw-Hill, Elsevier, John Wiley & Sons, ASME, SPE, and ASM International. Copyright: Knovel http://www.knovel.com/

Scitopia. Scitopia.org searches more than three and a half million documents, including peer-reviewed journal content and technical conference papers from leading voices in major science and technology disciplines. Searches bibliographic records in each partner's electronic library; patents from the U.S. Patent and Trademark Office, European Patent Office, and Japan Patent Office; and U.S. government documents on the Department of Energy Information Bridge site.

TRIS Online (1960 to the present). TRIS Online is a public-domain, Web-based version of the Transportation Research Information Services (TRIS) bibliographic database, as a component of the National Transportation Library. Its purpose is to enhance transportation research, safety, and operations by sharing knowledge and information.

Water Resources Abstracts (1967 to the present; Cambridge Scientific Abstracts). Provides summaries of the world's technical and scientific literature on water-related topics covering the characteristics, conservation, control, pollution, treatment, use, and management of water resources.

ENCYCLOPEDIA

Johnson, Stephen, and Roberto T. Leon. *Encyclopedia of Bridges and Tunnels.* New York: Facts on File, 2002. ISBN: 0816044821.

HANDBOOKS AND TABLES

American Institute of Timber Construction. *Timber Construction Manual,* 3rd ed. New York: Wiley, 1986. ISBN: 0471827584.

Barker, Richard M., and Jay A. Puckett. *Design of Highway Bridges: Based on AASHTO LRFD Bridge Design Specifications.* New York: John Wiley, 1997. ISBN: 0471304344.

Blake, L. S., ed. *Civil Engineer's Reference Book,* 4th ed. London: Butterworths, 1989. ISBN: 0408012080.

Brater, Ernest F., et al. *Handbook of Hydraulics,* 7th ed. Boston: McGraw-Hill, 1996. ISBN: 0070072477.

Breyer, Donald E., et al. *Design of Wood Structures ASD,* 4th ed. New York: McGraw-Hill, 1998. ISBN: 0070077169.

Bynum, Richard T. *Insulation Handbook.* New York: McGraw-Hill, 2001. ISBN: 0071348727.

Chen, Wai-Fah, and J.Y. Richard Liew, eds. *The Civil Engineering Handbook,* 2nd ed. Boca Raton, FL: CRC Press, 2003. ISBN: 0849309581.

Chen, Wai-Fah, and Lian Duan, eds. *Bridge Engineering Handbook.* Boca Raton, FL: CRC Press, 2000. ISBN: 0849374340.

Ching, Frank. *Building Construction Illustrated,* 4th ed. Hoboken, NJ: John Wiley & Sons, 2008. ISBN: 0470087811.

Dewberry, Sidney O., and Philip C. Champagne, eds. *Land Development Handbook: Planning, Engineering, and Surveying,* 2nd ed. New York: McGraw-Hill, 2002. ISBN: 0071375252.

Faherty, Keith F., and Thomas G. Williamson, eds. *Wood Engineering and Construction Handbook,* 3rd ed. New York: McGraw-Hill, 1997. ISBN: 0070220700.

Forest Products Laboratory. *Wood Engineering Handbook,* 2nd ed. Englewood Cliffs, NJ: Prentice Hall, 1990. ISBN: 0139637451.

Gaylord, Edwin H., et al., eds. *Structural Engineering Handbook,* 4th ed. New York: McGraw-Hill, 1997. ISBN: 0070237247.

Kreider, Jan F., ed. *Handbook of Heating, Ventilation, and Air Conditioning.* Boca Raton, FL: CRC Press, 2001. ISBN: 0849395844.

Kreith, Frank, and D. Yogi Goswami, eds. *Handbook of Energy Efficiency and Renewable Energy.* Boca Raton, FL: CRC Press, 2007. ISBN: 0849317304.

Kubba, Sam. *LEED Practices, Certification, and Accreditation Handbook.* Burlington, MA: Butterworth-Heinemann/Elsevier, 2010. ISBN: 1856176916.

Lamm, Ruediger, et al. *Highway Design and Traffic Safety Engineering Handbook.* New York: McGraw-Hill, 1999. ISBN: 0070382956.

Lincoln, William A. *World Woods in Color.* Fresno, CA: Linden Pub., 1991. ISBN: 0941936201.

Merritt, Frederick S., and Jonathan T. Ricketts, eds. *Building Design and Construction Handbook,* 6th ed. New York: McGraw-Hill, 2001. ISBN: 007041999X.

Nawy, Edward G, ed. *Concrete Construction Engineering Handbook.* Boca Raton, FL: CRC Press, 1997. ISBN: 0849326664.

Neville, Adam M. *Properties of Concrete.* 4th ed. New York: John Wiley & Sons, 1996. ISBN: 0470235276.

Nichols, Herbert L., Jr., and David A. Day. *Moving the Earth: The Workbook of Excavation*, 4th ed. New York: McGraw-Hill, 2005. ISBN: 0070464847.

Rowe, William H. *HVAC: Design Criteria, Options, Selection,* 2nd ed. Kingston, MA: R. S. Means Co., 1994. ISBN: 087629347X.

Scharff, Robert, and Terry Kennedy. *Roofing Handbook,* 2nd ed. New York: McGraw-Hill, 2001. ISBN: 0071360581.

Watson, Donald, ed. *Time-Saver Standards for Building Materials & Systems: Design Criteria and Selection Data.* New York: McGraw-Hill, 2000. ISBN: 0071356924.

WEB SITES

Bridge Engineering Home Page: http://www.scsolutions.com/bridge/

The Civil Engineer, a Center for Integrating Information on Civil Engineering: http://www.the civilengineer.org/

4 Specs.com: http://www.4specs.com/

iCivilEngineer: http://www.icivilengineer.com/. iCivilEngineer.com is a knowledge portal specially designed for civil engineering professionals and students. It has two goals in mind: (1) collect and catalog valuable civil engineering–relevant Internet resources so that people can find information fast; (2) explore how to take advantage of Internet technology to serve the civil engineering community.

Land Surveying Rules, Regulations, Statutes, and Minimum Standards: http://www.lsrp.com/statinfo.html

Multidisciplinary Center for Earthquake Engineering Research: http://mceer.buffalo.edu/

Soil Data Mart: http://soildatamart.nrcs.usda.gov/

TenLinks.com: Ultimate Civil Engineering Directory: http://www.tenlinks.com/engineering/civil/.

TenLinks, Inc., is an online CAD, CAM, and CAE media company. Founded in 1999 with TenLinks.com, TenLinks has grown into a network of Web sites and newsletters that together form the largest community of CAD, CAM, and CAE users on the Internet.

MISCELLANEOUS

Kavanagh, Barry F., and S. J. Glenn Bird. *Surveying: Principles and Applications*, 5th ed. Upper Saddle River, NJ: Prentice Hall, 2000. ISBN: 0130227331.

Plowden, David. *Bridges: The Spans of North America.* New York: Norton, 2002. ISBN: 0393050564.

Computer Science and Information Technology

BIBLIOGRAPHY

Cortado, J. W., comp. *Bibliographic Guide to the History of Computing: Computers and the Information Processing Industry.* Westport, CT: Greenwood Press, 1990. ISBN: 10031326810X.

DATABASES

AccessScience. This is the *McGraw-Hill Encyclopedia of Science and Technology* on the Web. It provides full access to articles, dictionary terms, and hundreds of research updates in all areas of science and technology.

ACM Digital Library (1985 to the present, coverage varies). This database contains ACM (Association for Computing Machinery) abstracts and full-text articles published in ACM journals, transactions, and proceedings.

Applied Science and Technology Abstracts (ASTA) (1983 to the present; abstracts 1994 to the present). Applied Science and Technology Abstracts covers core English-language scientific and technical publications. Topics include engineering, acoustics, chemistry, computers, metallurgy, physics, plastics, telecommunications, transportation, and waste management. Periodical coverage includes trade and industrial publications, journals issued by professional and technical societies, and specialized subject periodicals, as well as special issues such as buyers' guides, directories, and conference proceedings.

EI Compendex (also known as the Engineering Index) (1884 to the present). Compendex is the most comprehensive bibliographic database of scientific and technical engineering research available, covering all engineering disciplines. It includes millions of bibliographic citations and abstracts from thousands of engineering journals and conference proceedings. When combined with the Engineering Index Backfile (1884–1969), Compendex covers well over 120 years of core engineering literature.

IEEE Xplore (1988 to the present). Provides full-text access to IEEE transactions, journals, and conference proceedings and IEE journals from 1988 to the present, as well as all current IEEE standards. This database provides you with a single source leading to almost one-third of the world's current electrical engineering and computer science literature, with unparalleled access to IEEE and IEE publications.

INSPEC (selected earlier literature to 1896, 1898 to the present). Engineering Information, Inc. Contains literature in electrical engineering, electronics, physics, control engineering, information technology, communications, computers, computing, and manufacturing and production engineering. The database contains nearly 10 million bibliographic records taken from 3,850 scientific and technical journals and 2,200 conference proceedings. Approximately 330,000 new records are added to the database annually.

ISI Web of Science (1945 to the present, coverage varies). Institute for Scientific Information (ISI) http://scientific.thomson.com/isi/. A multidisciplinary database, with searchable author abstracts, covering the literature of the sciences, social sciences, and the arts and humanities. Users may choose to search across all the databases, or select just one. This unique database indexes and links cited references for each article. All formats provide complete bibliographic data and additional features, such as cited reference searching, links to related articles, plus author and publisher addresses.

Knovel: Answers for Science and Engineering (coverage varies). Knovel contains online interactive reference books and databases. It has a database of some of the leading science and engineering reference handbooks, databases, and conference proceedings from publishers such as McGraw-Hill, Elsevier, John Wiley & Sons, ASME, SPE, and ASM International. Copyright: Knovel http://www.knovel.com/

MathSciNet (1940 to the present; American Mathematical Society). A searchable database providing access to over 55 years of *Mathematical Reviews* and *Current Mathematical Publications*. MathSciNet contains over 2 million items and over 700,000 direct links to original articles in the mathematical sciences.

Scitopia. Scitopia.org searches more than three and a half million documents, including peer-reviewed journal content and technical conference papers from leading voices in major science and technology disciplines. Searches bibliographic records in each partner's electronic library; patents from the U.S. Patent and Trademark Office, European Patent Office, and Japan Patent Office; and U.S. government documents on the Department of Energy Information Bridge site.

U.S. Patent and Trademark Office Patent Databases (1790 to the present). These databases (http://www.uspto.gov/) allow users to search and view the full-text contents and bibliographic information of patents. Users can search by multiple elements of patent records such as patent numbers, titles, inventors, application dates, descriptions/specifications, and so on. The USPTO houses full text for patents issued from 1976 to the present and TIFF images for all patents from 1790 to the present.

DICTIONARIES

Daintith, John. *The Facts on File Dictionary of Computer Science.* Facts on File, c2006. ISBN: 0816059993.

Laplante, Phillip A. *Dictionary of Computer Science, Engineering, and Technology.* Boca Raton, FL: CRC Press, c2001. ISBN: 0849326915.

McGraw-Hill Book Company. *Dictionary of Computing and Communications.* New York: McGraw-Hill, 2003. ISBN: 0071421785.

ENCYCLOPEDIAS

Encyclopedia of Computer Science. London: Nature Publishing Group, 2000. ISBN: 0333778790.

Fitzroy Dearborn. *Encyclopedia of 20th-Century Technology.* New York: Routledge, 2005. ISBN: 101579584632.

Gibilisco, Stan, ed. *Concise Encyclopedia of Robotics.* New York: McGraw-Hill, 2003. ISBN: 0071410104.

Gibilisco, Stan, ed. *The McGraw-Hill Illustrated Encyclopedia of Robotics & Artificial Intelligence.* New York: McGraw-Hill, 1994. ISBN: 0070236135.

HANDBOOK AND TABLE

Levine, William S., ed. *The Control Handbook.* Boca Raton, FL: CRC Press, 1996. ISBN: 0849385709.

WEB SITES

CiteSeerX, Scientific Literature Digital Library and Search Engine: http://citeseerx.ist.psu.edu/

Computer Oral History Project of the Smithsonian Museum of American History, Computer History Collection: http://americanhistory.si.edu/collections/comphist/

The Turing Archive for the History of Computing: http://www.alanturing.net/

MISCELLANEOUS

Agar, J. *Turing and the Universal Machine: The Making of the Modern Computer.* Lanham, MD: Totem Books, 1997. ISBN: 101840462507.

Dubbey, J. M. *The Mathematical Work of Charles Babbage.* Cambridge: Cambridge University Press, 1978. ISBN: 100521524768.

Williams, M. R. *A History of Computing Technology,* 2nd ed. Los Alamitos, CA: IEEE Computer Society Press, 1997. ISBN: 100818677392.

Electrical Engineering

DICTIONARIES

Institute of Electrical and Electronics Engineers. *The IEEE Standard Dictionary of Electrical and Electronics Terms,* 6th ed. New York: Institute of Electrical and Electronics Engineers, 1996. ISBN: 1559378336.

Kaplan, Steven M. *Wiley Electrical and Electronics Engineering Dictionary.* Hoboken, NJ: Wiley-Interscience, 2004. ISBN: 0471402249.

Petersen, Julie K. *Fiber Optics Illustrated Dictionary.* Boca Raton, FL: CRC Press, c2003. ISBN: 084931349X.

Titus, Patricia A., et al. *Illustrated Dictionary for Electrical Workers,* 2nd ed. Albany, NY: Delmar/Thomson Learning, 2002. ISBN: 0766828530.

White, Glenn D., and Gary J. Louie. *The Audio Dictionary, 3rd rev.* Seattle: University of Washington Press, c2005. ISBN: 0295984988.

DATABASES

AccessScience. This is the *McGraw-Hill Encyclopedia of Science and Technology* on the Web. It provides full access to articles, dictionary terms, and hundreds of research updates in all areas of science and technology.

From *Science and Technology Resources: A Guide for Information Professionals and Researchers* by James E. Bobick and G. Lynn Berard. Santa Barbara, CA: Libraries Unlimited. Copyright © 2011.

ACM Digital Library (1985 to the present, coverage varies). This database contains ACM (Association for Computing Machinery) abstracts and full-text articles published in ACM journals, transactions, and proceedings.

Applied Science and Technology Abstracts (ASTA) (1983 to the present; abstracts 1994 to the present). Applied Science and Technology Abstracts covers core English-language scientific and technical publications. Topics include engineering, acoustics, chemistry, computers, metallurgy, physics, plastics, telecommunications, transportation, and waste management. Periodical coverage includes trade and industrial publications, journals issued by professional and technical societies, and specialized subject periodicals, as well as special issues such as buyers' guides, directories, and conference proceedings.

EI Compendex (also known as the Engineering Index) (1884 to the present). Compendex is the most comprehensive bibliographic database of scientific and technical engineering research available, covering all engineering disciplines. It includes millions of bibliographic citations and abstracts from thousands of engineering journals and conference proceedings. When combined with the Engineering Index Backfile (1884–1969), Compendex covers well over 120 years of core engineering literature.

IEEE Xplore (1988 to the present). Provides full-text access to IEEE transactions, journals, and conference proceedings and IEE journals from 1988 to the present, as well as all current IEEE standards. This database provides you with a single source leading to almost one-third of the world's current electrical engineering and computer science literature, with unparalleled access to IEEE and IEE publications.

INSPEC (selected earlier literature to 1896, 1898 to the present). Engineering Information, Inc. Contains literature in electrical engineering, electronics, physics, control engineering, information technology, communications, computers, computing, and manufacturing and production engineering. The database contains nearly 10 million bibliographic records taken from 3,850 scientific and technical journals and 2,200 conference proceedings. Approximately 330,000 new records are added to the database annually.

ISI Web of Science (1945 to the present, coverage varies). Institute for Scientific Information (ISI) http://scientific.thomson.com/isi/. A multidisciplinary database, with searchable author abstracts, covering the literature of the sciences, social sciences, and the arts and humanities. Users may choose to search across all the databases, or select just one. This unique database indexes and links cited references for each article. All formats provide complete bibliographic data and additional features, such as cited reference searching, links to related articles, plus author and publisher addresses.

Knovel: Answers for Science and Engineering (coverage varies). Knovel contains online interactive reference books and databases. It has a database of some of the leading science and engineering reference handbooks, databases, and conference proceedings from publishers such as McGraw-Hill, Elsevier, John Wiley & Sons, ASME, SPE, and ASM International. Copyright: Knovel http://www.knovel.com/

Scitopia. Scitopia.org searches more than three and a half million documents, including peer-reviewed journal content and technical conference papers from leading voices in major science and technology disciplines. Searches bibliographic records in each partner's electronic library; patents from the U.S. Patent and Trademark Office, European Patent Office, and Japan Patent Office; and U.S. government documents on the Department of Energy Information Bridge site.

U.S. Patent and Trademark Office Patent Databases (1790 to the present). These databases (http://www.uspto.gov/) allow users to search and view the full-text contents and bibliographic information of patents. Users can search by multiple elements of patent records such as patent numbers, titles, inventors, application dates, descriptions/specifications, and so on. The USPTO houses full text for patents issued from 1976 to the present and TIFF images for all patents from 1790 to the present.

ENCYCLOPEDIAS

McGraw-Hill Concise Encyclopedia of Science and Technology. New York: McGraw-Hill, c2005. ISBN: 007149573.

Muller, Nathan J. *Desktop Encyclopedia of Telecommunications.* New York: McGraw-Hill, 2002. ISBN: 0071381481.

Webster, John G., ed. *Wiley Encyclopedia of Electrical and Electronics Engineering.* New York: John Wiley, 1999. ISBN: 0471139467.

HANDBOOKS AND TABLES

Beaty, H. Wayne, and James L. Kirtley Jr. *Electric Motor Handbook.* New York: McGraw-Hill, 1998. ISBN: 0070359717.

Dorf, Richard C., ed. *The Electrical Engineering Handbook*, 2nd ed. Boca Raton, FL: CRC Press, 1997. ISBN: 0849385741.

Elliott, Thomas C., ed. *Standard Handbook of Powerplant Engineering.* New York: McGraw-Hill, 1989. ISBN: 0070191069.

Grigsby, Leonard L., ed. *The Electric Power Engineering Handbook.* Boca Raton, FL: CRC Press, 2001. ISBN: 0849385784.

McPartland, Joseph F., et al. *McGraw-Hill's Handbook of Electrical Construction Calculations.* Rev. ed. New York: McGraw-Hill, 1998. ISBN: 0070466416.

Rea, Mark S., ed. *The IESNA Lighting Handbook: Reference & Application*, 9th ed. New York: Illuminating Engineering Society of North America, 2000. ISBN: 0879951508.

Shoemaker, Thomas M., and James E. Mack. *The Lineman's and Cableman's Handbook,* 10th ed. New York: McGraw-Hill, 2002. ISBN: 0071362401.

WEB SITES

Electronics Weekly: http://www.electronicsweekly.com/Home/

The Engineer: http://www.theengineer.co.uk/sectors/electronics/

GlobalSpec: The Engineering Search Engine: http://www.globalspec.com/

Institute of Electrical and Electronics Engineers: http://www.ieee.org/index.html

Environmental Engineering

HISTORY

Peter S. Thacher Environment Collection, 1960–1996. Environmental Science and Public Policy Archives, Harvard College. http://oasis.lib.harvard.edu/oasis/deliver/~env00003

DICTIONARIES

American Water Works Association. *The Drinking Wter Dctionary,* 2nd ed. American Water Works Association, 2005. ISBN: 1583213449.

Dauvergne, Peter. *Historical Dictionary of Environmentalism.* Lanham, MD: Scarecrow Press, 2009. ISBN: 9780810858046.

Lee, C.C., ed. *Environmental Engineering Dictionary,* 2nd ed. Rockville, MD: Government Institutes, 1992. ISBN: 0865873283.

Pankratz, Thomas M. *Environmental Engineering Dictionary and Directory.* Boca Raton, FL: Lewis, 2001. ISBN: 1566705436.

Smith, Paul G., and John S. Scott. *Dictionary of Water and Waste Management,* 2nd ed. Boston: Butterworth-Heinemann, 2002. ISBN:0750646381.

Webster, Len F. *A Dictionary of Environmental & Civil Engineering.* New York: Parthenon Pub. Group, 2000. ISBN: 1850700753.

DATABASES

Applied Science and Technology Abstracts (ASTA) (1983 to the present; abstracts 1994 to the present). Applied Science and Technology Abstracts covers core English-language scientific and technical publications. Topics include engineering, acoustics, chemistry, computers, metallurgy, physics, plastics, telecommunications, transportation, and waste management. Periodical coverage includes trade and industrial publications, journals issued by professional and technical societies, and specialized subject periodicals, as well as special issues such as buyers' guides, directories, and conference proceedings.

BuildingGreen Suite. The BuildingGreen Suite brings together articles, reviews, and news from *Environmental Building News* (*EBN*) since 1992, product listings from the Green-Spec products directory, and project case studies from the high-performance buildings database.

CSA Illustrata: Natural Sciences (1977 to the present). This Cambridge Scientific Abstracts database indexes tables, figures, graphs, charts, and other illustrations from scholarly literature in agriculture, biology, conservation, earth sciences, environmental studies, fish and fisheries, food and food industries, forests and forestry, geography, medical sciences, meteorology, veterinary science, and water resources since 1997.

EI Compendex (also known as the Engineering Index) (1884 to the present). Compendex is the most comprehensive bibliographic database of scientific and technical engineering research available, covering all engineering disciplines. It includes millions of bibliographic citations and abstracts from thousands of engineering journals and conference proceedings. When combined with the Engineering Index Backfile (1884–1969), Compendex covers well over 120 years of core engineering literature.

Environmental Sciences and Pollution Management (1967 to the present). This multidisciplinary database, provides unparalleled and comprehensive coverage of the environmental sciences. Abstracts and citations are drawn from over 10,000 serials including scientific journals, conference proceedings, reports, monographs, books, and government publications.

ISI Web of Science (1945 to the present, coverage varies). Institute for Scientific Information (ISI) http://scientific.thomson.com/isi/. A multidisciplinary database, with searchable author abstracts, covering the literature of the sciences, social sciences, and the arts and humanities. Users may choose to search across all the databases, or select just one. This unique database indexes and links cited references for each article. All formats provide complete bibliographic data and additional features, such as cited reference searching, links to related articles, plus author and publisher addresses.

Knovel: Answers for Science and Engineering (coverage varies). Knovel contains online interactive reference books and databases. It has a database of some of the leading science and engineering reference handbooks, databases, and conference proceedings from publishers such as McGraw-Hill, Elsevier, John Wiley & Sons, ASME, SPE, and ASM International. Copyright: Knovel http://www.knovel.com/

LexisNexis Congressional. This site includes congressional publications, current topics of debate, members and committees, full text of proposed and recent bills, laws and regulations, and much more. From this site, you can also search the *Federal Register* and other documents.

National Library of the Environment. This is a free, non-advocacy online environmental information service managed by the Committee for the National Institute for the Environment

(CNIE), a non-profit organization (http://www.ncseonline.org/NLE/). CNIE's mission is to improve the scientific basis for making decisions on environmental issues through the successful operation of a National Institute for the Environment (NIE).

SciFinder Scholar (1907 to the present; also known as Chemical Abstracts). Search millions of chemical references and substances by research topic, author name, structure and substructure, chemical name or CAS Registry Number, company/name organization, as well as browse the tables of contents in journals, and enable a reaction query.

THOMAS. In the spirit of Thomas Jefferson, legislative information from the Library of Congress can be found at this site (http://thomas.loc.gov/home/abt_thom.html). The leadership of the 104th Congress directed the Library of Congress to make federal legislative information freely available to the public. Search for legislation, committees and their reports, congressional voting, presidential nominations, and so on.

U.S. GOVERNMENT POLICY WEB SITES/DATABASES

For legislation in the House:

http://energycommerce.house.gov/Subcommittees/ehm.shtml

http://resourcescommittee.house.gov/index.php

For legislation in the Senate:

http://energy.senate.gov/public/

http://energy.senate.gov/public/index.cfm?FuseAction=About.Subcommittee&Subcom mittee_ID=0ee9d4b8-cb23-42ee-ac0f-fc7c6897855f

http://epw.senate.gov/public/

http://epw.senate.gov/public/index.cfm?FuseAction=Subcommittees.Subcommittee& Subcommittee_id=47912570-0861-4d66-9490-fe0f98304abf

ENCYCLOPEDIAS

Kirk-Othmer Encyclopedia of Chemical Technology, 4th ed. [electronic resource]. New York: Wiley, 2000.

Ullmann's Encyclopedia of Industrial Chemistry, 6th ed. [electronic resource]. New York: Wiley, c2000.

Water Encyclopedia [electronic resource]. Jay H. Lehr. Hoboken, NJ: John Wiley & Sons, c2005.

HANDBOOKS AND TABLES

Allport, Dennis C. *MDI and TDI: A Safety, Health and the Environment: A Source Book and Practical Guide.* New York: J. Wiley, c2003. ASIN: B004A16MIY.

Corbitt, Robert A., ed. *Standard Handbook of Environmental Engineering,* 2nd ed. New York: McGraw-Hill, 1999. ISBN: 0070131600.

Eaton, Andrew D. *Standard Methods for the Examination of Water & Wastewater,* 21st ed. Washington, DC: American Public Health Association, c2005. ISBN: 9780875530475.

Liu, David H. F., and Béla G. Lipták, eds. *Environmental Engineers' Handbook,* 2nd ed. Boca Raton, FL: Lewis Publishers, 1997. ISBN: 0849399718.

Reynolds, Joseph P. *Handbook of Chemical and Environmental Engineering Calculations.* New York: J. Wiley, c2002. ISBN: 1591247403.

Thai, Khi V., Dianne Rahm, and Jerrell D. Coggburn, eds. *Handbook of Globalization and the Environment.* Boca Raton, FL: CRC Press, c2007. ISBN: 1574445537.

DIRECTORIES

The Electronic Code of Federal Regulations (e-CFR). It is a currently updated version of the Code of Federal Regulations (CFR) (http://ecfr.gpoaccess.gov/cgi/t/text/text-idx?c=ecfr&tpl=%2Findex.tpl). It is not an official legal edition of the CFR.

The Federal Register. Published by the Office of the Federal Register, National Archives and Records Administration (NARA) (http://www.gpoaccess.gov/fr/), the *Federal Register* is the official daily publication for rules, proposed rules, and notices of federal agencies and organizations, as well as executive orders and other presidential documents.

WEB SITES

The National Library for the Environment (NLE) currently posts 2050 CRS Reports on environmental and related topics. The Congressional Research Service (CRS), part of the Library of Congress, prepares its reports for the U.S. Congress. CRS products undergo review for accuracy and objectivity and contain nontechnical information that can be very useful to people interested in environmental policy: http://ncseonline.org/NLE/CRs/

Regulations.gov is your source for all regulations (or rule makings) issued by U.S. government agencies: http://www.regulations.gov/search/index.jsp

MISCELLANEOUS

CQ Weekly. Here (http://library.cqpress.com/cqweekly/) readers get in-depth reports on issues looming on the congressional horizon, plus a complete wrap up of the previous week's news, including the status of bills in play, behind-the-scenes maneuvering, committee and floor activity, debates and all roll-call votes.

Environmental Science

HISTORIES

Brooks, Paul. *Speaking for Nature: How Literary Naturalists from Henry Thoreau to Rachel Carson Have Shaped America.* Boston: Houghton Mifflin Co., 1980. ISBN: 0395296102.

McKibben, Bill, ed. *American Earth: Environmental Writing since Thoreau.* New York: Literary Classics of the United States, 2008. ISBN: 1598530208.

DICTIONARIES

Bailey, Jill, ed. *The Facts on File Dictionary of Ecology and the Environment.* New York: Facts on File, 2004. ISBN: 081604922X.

Environmental Health & Safety Dictionary: Official Regulatory Terms, 7th ed. Rockville, MD: Government Institutes, 2000. ISBN: 0865876886.

Lewis, Robert A. *Lewis' Dictionary of Toxicology.* Boca Raton, FL: Lewis Publishers, 1998. ISBN: 1566702232.

Wyman, Bruce C., and L. Harold Stevenson. *The Facts on File Dictionary of Environmental Science.* New York: Checkmark Books, 2001. ISBN: 0816042330.

DATABASES

CSA Illustrata: Natural Sciences (1977 to the present). This Cambridge Scientific Abstracts database indexes tables, figures, graphs, charts, and other illustrations from scholarly literature in agriculture, biology, conservation, earth sciences, environmental studies, fish and fisheries, food and food industries, forests and forestry, geography, medical sciences, meteorology, veterinary science, and water resources since 1997.

Environmental Sciences and Pollution Management (1967 to the present). This multidisciplinary database, provides unparalleled and comprehensive coverage of the environmental sciences. Abstracts and citations are drawn from over 10,000 serials including scientific journals, conference proceedings, reports, monographs, books, and government publications.

GeoBase (1980 to the present; Elsevier, Inc.). GeoBase is a multidisciplinary database covering human and physical geography, geology, oceanography, geomechanics, alternative energy sources, pollution, waste management, and nature conservation.

ISI Web of Science (1945 to the present, coverage varies). Institute for Scientific Information (ISI) http://scientific.thomson.com/isi/. A multidisciplinary database, with searchable author abstracts, covering the literature of the sciences, social sciences, and the arts and humanities. Users may choose to search across all the databases, or select just one. This unique database indexes and links cited references for each article. All formats provide complete bibliographic data and additional features, such as cited reference searching, links to related articles, plus author and publisher addresses.

National Library of the Environment. This is a free, non-advocacy online environmental information service managed by the Committee for the National Institute for the Environment (CNIE), a non-profit organization (http://www.ncseonline.org/NLE/). CNIE's mission is to improve the scientific basis for making decisions on environmental issues through the successful operation of a National Institute for the Environment (NIE).

ENCYCLOPEDIAS

Allaby, Michael. *Biomes of the World.* Danbury, CT: Grolier Educational, 1999. 9 vols. ISBN: 0717293416.

Ashworth, William, and Charles E. Little. *Encyclopedia of Environmental Studies.* New York: Facts on File, 2001. ISBN: 0816042551.

Benke, Arthur C., and Colbert E. Cushing, eds. *Rivers of North America.* Boston: Elsevier/Academic Press, 2005. ISBN: 0120882531.

Benson, Sonia, and Rob Nagel. *Endangered Species*, 2nd ed. Detroit: UXL, 2003. 3 vols. ISBN: 0787676187.

Bingham, Eula, et al., eds. *Patty's Toxicology*, 5th ed. New York: John Wiley & Sons, 2001. 9 vols. ISBN: 0471319430.

Bisio, Attilio, and Sharon Boots, eds. *The Wiley Encyclopedia of Energy and the Environment.* New York: Wiley, 1996. 2 vols. ISBN: 47114827X.

Cleveland, Cutler. J., ed. *Encyclopedia of Energy.* Boston, MA: Elsevier Academic Press, 2004. 6 vols. ISBN: 012176480X.

Crawford, Mark. *Toxic Waste Sites: An Encyclopedia of Endangered America.* Santa Barbara, CA: ABC-CLIO, 1997. ISBN: 0874369347.

Crawford, Mark. *Habitats and Ecosystems: An Encyclopedia of Endangered America.* Santa Barbara, CA: ABC-CLIO, 1999. ISBN: 0874369975.

Dutch, Steven I., ed. *Encyclopedia of Global Warming.* Pasadena, CA: Salem Press, 2010. 3 vols. ISBN: 1587655632.

Editors of Salem Press. *Ecology Basics.* Pasadena, CA: Salem Press, 2004. 2 vols. ISBN: 1587651742.

Encyclopedia of Endangered Species. Detroit: Gale Research, 1994. 2 vols. ISSN: 10771352.

Environmental Viewpoints: Selected Essays and Excerpts on Issues in Environmental Protection. Detroit: Gale Research, Inc. 3 vols. ISSN: 1063116X.

Greenberg, Michael I. et al., eds. *Occupational, Industrial, and Environmental Toxicology*, 2nd ed. Philadelphia: Mosby, 2003. ISBN: 0323013406.

Jukofsky, Diane. *Encyclopedia of Rainforests.* Westport, CT: Oryx Press, 2002. ISBN: 1573562599.

Klaassen, Curtis D., et al., eds. *Casarett and Doull's Toxicology: The Basic Science of Poisons*, 7th ed. New York: McGraw-Hill, 2008. ISBN: 0071470514.

Levin, Simon A., ed. *Encyclopedia of Biodiversity.* San Diego: Academic Press, 2001. 5 vols. ISBN: 0122268652.

Marshall Cavendish Corporation. *Endangered Wildlife and Plants of the World.* New York: Marshall Cavendish, 2001. 13 vols. ISBN: 0761471944.

Nierenberg, William A., ed. *Encyclopedia of Environmental Biology.* San Diego: Academic Press, 1995. 3 vols. ISBN: 0122267303.

Stapleton, Richard M., ed. *Pollution A to Z.* New York: Macmillan Reference USA: Thomson/Gale, 2004. 2 vols. ISBN: 0028657004.

Stewart, Bobby A., and Terry A. Howell, eds. *Encyclopedia of Water Science.* New York: Marcel Dekker, 2003. ISBN: 0824709489.

Turkington, Carol, and Deborah Mitchell. *The Encyclopedia of Poisons and Antidotes*, 3rd ed. New York: Facts on File, 2010. ISBN: 9780816064014.

Van der Leeden, Frits, et al. *The Water Encyclopedia*, 2nd ed. Chelsea, MI: Lewis Publishers, 1990. ISBN: 0873711203.

Weigel, Marlene. *U-X-L Encyclopedia of Biomes.* Detroit: U-X-L, 2000. 3 vols. ISBN: 0787637327.

Wexler, Philip, and Shayne C. Gad, eds. *Encyclopedia of Toxicology.* San Diego: Academic Press, 1998. 3 vols. ISBN: 012227220X.

Zumerchik, John, ed. *Macmillan Encyclopedia of Energy.* New York: Macmillan Reference USA, 2001. 3 vols. ISBN: 0028650212.

HANDBOOKS AND TABLES

Beacham, Walton, ed. *World Wildlife Fund Guide to Extinct Species of Modern Times.* Osprey, FL: Beacham Pub., 1997. ISBN: 0933833407.

Beacham, Walton, and Kirk H. Beetz, eds. *Beacham's Guide to International Endangered Species.* Osprey, FL: Beacham Pub., 1998. 3 vols. ISBN: 0933833342.

Beacham, Walton, et al., eds. *Beacham's Guide to the Endangered Species of North America.* Detroit: Gale Group, 2001. 6 vols. ISBN: 0787650285.

Bevelacqua, Armando S., and Richard H. Stilp. *Hazardous Materials Field Guide*, 2nd ed. Florence, KY: Delmar Cengage Learning, 2006. ISBN: 1418038288.

Book of Lists for Regulated Hazardous Substances, 9th ed. Rockville, MD: Government Institutes, 2010. ISSN: 1521-4427.

Briggs, Shirley A. *Basic Guide to Pesticides: Their Characteristics and Hazards.* Washington: Hemisphere Pub. Corp., 1992. ISBN: 1560322535.

Burke, Gwendolyn, et al. *Handbook of Environmental Management and Technology.* New York: Wiley, 1993. ISBN: 047158584X.

Derelanko, Michael J., and Mannfred A. Hollinger, eds. *CRC Handbook of Toxicology.* Boca Raton, FL: CRC Press, 1995. ISBN: 0849386683.

DHHS Centers for Disease Control. *NIOSH Pocket Guide to Chemical Hazards.* Baton Rogue, LA: Claitor's Law Books and Publishing Division, 2005. ISBN: 1598040529.

Freeman, Harry M., ed. *Standard Handbook of Hazardous Waste Treatment and Disposal.* New York: McGraw-Hill, 1989. ISBN: 0070220425.

Genium's Handbook of Safety, Health, and Environmental Data for Common Hazardous Substances. New York: McGraw Hill, 1998. 3 vols. ISBN: 0079136141.

Keith, Lawrence H., and Mary Walker. *Handbook of Air Toxics: Sampling, Analysis, and Properties.* Boca Raton, FL: CRC Press, 1995. ISBN: 1566701147.

Kreith, Frank, ed. *Handbook of Solid Waste Management.* New York: McGraw-Hill, 1994. ISBN: 0070358761.

Lewis, Richard J. *Hazardous Chemicals Desk Reference*, 6th ed. Hoboken, NJ: Wiley-Interscience, 2008. ISBN: 0470180242.

Lewis, Richard J., Sr., and N. Irving Sax. *Sax's Dangerous Properties of Industrial Materials,* 11th ed. Hoboken, NJ: Wiley-Interscience, 2011. 3 vols. ISBN: 0471476625.

Lund, Herbert F. *The McGraw-Hill Recycling Handbook*, 2nd ed. New York: McGraw-Hill, 2001. ISBN: 0070391564.

Middleton, Susan, and David Liittschwager. *Witness: Endangered Species of North America.* San Francisco: Chronicle Books, 1994. ISBN: 0811802825.

Miller, E. Willard, and Ruby M. Miller. *Indoor Pollution: A Reference Handbook.* Santa Barbara, CA: ABC-CLIO, 1998. ISBN: 0874368952.

Milne, George W. A., ed. *Pesticides: An International Guide to 1,800 Pest Control Chemicals,* 2nd ed. Burlington, VT: Ashgate, 2004. ISBN: 0566085429.

Moan, Jaina L., and Zachary A. Smith. *Energy Use Worldwide: A Reference Handbook.* Santa Barbara, CA: ABC-CLIO, 2007. ISBN: 9781851098903.

Montgomery, John H. *Agrochemicals Desk Reference,* 2nd ed. Boca Raton, FL: CRC Press, 1997. ISBN: 1566701678.

Patnaik, Pradyot. *A Comprehensive Guide to the Hazardous Properties of Chemicals Substances,* 3rd ed. Hoboken, NJ: Wiley-Interscience, 2007. ISBN: 0471714585.

Pohanish, Richard. *Sittig's Handbook of Toxic and Hazardous Chemicals and Carcinogens,* 5th ed. Norwich, NY: William Andrew Publishing, 2008. 2 vols. ISBN: 0815515537.

Pohanish, Richard P., and Stanley A. Greene, eds. *Hazardous Substances Resource Guide*, 2nd ed. Detroit: Gale, 1997. ISBN: 0810390620.

Spero, Jennifer, et al. *Regulatory Chemicals Handbook.* New York: Marcel Dekker, 2000. ISBN: 0824703901.

Strong, Debra L. *Recycling in America: A Reference Handbook*, 2nd ed. Santa Barbara, CA: ABC-CLIO, 1997. ISBN: 0874368898.

True, Bev-Lorraine, and Robert H. Dreisbach, eds. *Dreisbach's Handbook of Poisoning: Prevention, Diagnosis and Treatment*, 13th ed. London: Informa Healthcare, 2001. ISBN: 1850700389.

Urben, Peter. *Bretherick's Handbook of Reactive Chemical Hazards*, 7th ed. San Diego: Academic Press, 2006. 2 vols. ISBN: 0123725631.

Verschueren, Karel. *Handbook of Environmental Data on Organic Chemicals*, 4th ed. New York: Wiley, 2001. 2 vols. ISBN: 0471374903.

Vickers, Amy. *Handbook of Water Use and Conservation.* Amherst, MA: Waterplow Press, 2001. ISBN: 1931579075.

Woodside, Gayle. *Hazardous Materials and Hazardous Waste Management*, 2nd ed. New York: John Wiley & Sons, 1999. ISBN: 0471174491.

ATLASES

Burton, John A., ed. *The Atlas of Endangered Species,* 2nd ed. New York: Macmillan Library Reference, 1999. ISBN: 0028650344.

Groombridge, Brian, and Martin D. Jenkins. *World Atlas of Biodiversity: Earth's Living Resources in the 21st Century.* Berkeley: University of California Press, 2002. ISBN: 0520236688.

Mackay, Richard. *The Atlas of Endangered Species.* Berkeley: University of California Press, 2009. ISBN: 9780520258624.

WEB SITES

CDC's National Center for Environmental Health (NCEH): http://www.cdc.gov/nceh/

Environmental Protection Agency: http://www.epa.gov/

EPA Science Inventory: http://cfpub.epa.gov/si/

National Archives ARC Gallery, Environment: http://www.archives.gov/research/arc/topics/environment/

Natural Resources Conservation Service: http://www.nrcs.usda.gov/

MISCELLANEOUS

Bregman, Jacob I. *Environmental Impact Statements,* 2nd ed. Boca Raton, FL: Lewis Publishers, 1999. ISBN: 1566703697.

Flannery, Tim F., and Peter Schouten. *A Gap in Nature: Discovering the World's Extinct Animals.* New York: Atlantic Monthly Press, 2001. ISBN: 0871137976.

General Biology and Natural History

GUIDES TO THE LITERATURE

Bazler, Judith. *Biology Resources in the Electronic Age.* Westport, CT: Greenwood Press, 2003. ISBN: 1573563803.

Schmidt, Diane. *A Guide to Field Guides: Identifying the Natural History of North America.* Englewood, CO: Libraries Unlimited, 1999. ISBN: 1563087073.

Schmidt, Diane, et al. *Using the Biological Literature: A Practical Guide.* New York: Marcel Dekker, 2002. ISBN: 0824706676.

HISTORIES

Magner, Lois N. *A History of the Life Sciences,* 3rd ed. New York: Marcel Dekker, 2002. ISBN: 0824708245.

Pauly, Philip J. *Biologists and the Promise of American Life: From Meriwether Lewis to Alfred Kinsey.* Princeton, NJ: Princeton University Press, 2000. ISBN: 0691049777.

Singer, Charles J. *A History of Biology to about the Year 1900: A General Introduction to the Study of Living Things.* Ames: Iowa State University Press, 1989. ISBN: 0813809371.

BIBLIOGRAPHY

Overmier, Judith A. *The History of Biology: A Selected, Annotated Bibliography.* New York: Garland Pub., 1989. ISBN: 0824091183.

DICTIONARIES

Allaby, Michael, ed. *The Oxford Dictionary of Natural History.* Oxford: Oxford University Press, 1985. ISBN: 0192177206.

A Dictionary of Biology. Oxford: Oxford University Press, 2004. ISBN: 0198609175.

Hancock, John M., and Marketa J. Zvelebil. *Dictionary of Bioinformatics and Computational Biology.* Hoboken, NJ: Wiley-Liss, 2004. ISBN: 0471436224.

Hine, Robert, ed. *The Facts on File Dictionary of Biology*, 3rd ed. New York: Facts on File, 1999. ISBN: 0816039070.

Hine, Robert, ed. *Oxford Dictionary of Biology*, 6th ed. New York: Oxford University Press, 2008. ISBN: 0199204624.

Indge, Bill. *Dictionary of Biology.* London: Fitzroy Dearborn Publishers, 1999. ISBN: 1579581285.

Lawrence, Eleanor, ed. *Henderson's Dictionary of Biology*, 13th ed. New York: Pearson/Prentice Hall, 2005. ISBN: 0131273841.

ELECTRONIC DATABASES

Applied Science and Technology Abstracts (ASTA) (1983 to the present; abstracts 1994 to the present). Applied Science and Technology Abstracts covers core English-language scientific and technical publications. Topics include engineering, acoustics, chemistry, computers, metallurgy, physics, plastics, telecommunications, transportation, and waste management. Periodical coverage includes trade and industrial publications, journals issued by professional and technical societies, and specialized subject periodicals, as well as special issues such as buyers' guides, directories, and conference proceedings.

Biosis Previews (1926 to the present). Biosis Previews is a key database for literature searching in biological and life sciences. Covers medical topics with focus on basic research and its coverage of clinical medicine.

Chemical Abstracts (1907 to the present; available electronically via SciFinder Scholar database). SciFinder Scholar is an interface to the *Chemical Abstracts Service* (*CAS*). It allows you to search millions of chemical references and substances by research topic, author name, structure and substructure, chemical name or CAS Registry Number, company/name organization, as well as browse the tables of contents in journals, and enable a reaction query.

HSTM: History of Science, Technology, and Medicine (1975 to the present). Contains references to journal articles, chapters, and reviews within the fields of general science, technology, and medicine. It covers the influence of these fields on society and culture from prehistory to the present.

ISI Web of Science (1945 to the present, coverage varies). Institute for Scientific Information (ISI) http://scientific.thomson.com/isi/. A multidisciplinary database, with searchable

author abstracts, covering the literature of the sciences, social sciences, and the arts and humanities. Users may choose to search across all the databases, or select just one. This unique database indexes and links cited references for each article. All formats provide complete bibliographic data and additional features, such as cited reference searching, links to related articles, plus author and publisher addresses.

ENCYCLOPEDIAS

Baltimore, David, et al., eds. *Frontiers of Life.* San Diego, CA: Academic Press, 2002. 4 vols. ISBN: 0120773406.

Bruno, Leonard C., and Julie L. Carnagie, eds. *U.X.L. Complete Life Science Resource.* Detroit, MI: U.X.L., 2001. 3 vols. ISBN: 0787648515.

McGrath, Kimberley A., ed. *World of Biology.* Detroit, MI: Gale Group, 1999. ISBN: 0787630446.

Nature Encyclopedia. New York: Oxford University Press, 2001. ISBN: 0195218345.

O'Daly, Anne, ed. *Encyclopedia of Life Sciences*, 2nd ed. Tarrytown, NY: Marshall Cavendish, 2004. 13 vols. ISBN: 0761474420.

Post, Stephen G., ed. *Encyclopedia of Bioethics*, 3rd ed. New York: MacMillan Reference USA: Thomson/Gale, 2004. 5 vols. ISBN: 0028657748.

Rittner, Don, and Timothy L. McCabe. *Encyclopedia of Biology.* New York: Facts on File, 2004. ISBN: 0816048592.

Robinson, Richard, ed. *Biology.* New York: MacMillan Reference USA: Gale Group/Thomson Learning, 2002. 4 vols. ISBN: 0028655516.

Scott, Thomas A. *Concise Encyclopedia Biology.* Berlin: Walter de Gruyter, 1996. ISBN: 3110106612.

HANDBOOKS AND TABLES

Altman, Philip L., and Dorothy S. Dittmer, eds. *Biology Data Book*, 2nd ed. Bethesda, MD: Federation of American Societies for Experimental Biology, 1972 to the present. 3 vols. ISBN: 0913822078.

Barnes-Svarney, Patricia L., and Thomas E. Svarney. *The Oryx Guide to Natural History: The Earth and All Its Inhabitants.* Phoenix, AZ: Oryx Press, 1999. ISBN: 1573561592.

Clewis, Beth. *Index to Illustrations of Animals and Plants.* New York: Neal-Schuman, 1991. ISBN: 1555700721.

Council of Biology Editors. *Scientific Style and Format: The CBE Manual for Authors, Editors, and Publishers*, 6th ed. Cambridge: Cambridge University Press, 1994. ISBN: 0521471540.

Kotyk, Arnost. *Quantities, Symbols, Units, and Abbreviations in the Life Sciences: A Guide for Authors and Editors.* Totowa, NJ. Humana Press, 1999. ISBN: 0896036499.

Margulis, Lynn, and Karlene V. Schwartz. *Five Kingdoms: An Illustrated Guide to the Phyla of Life on Earth*, 3rd ed. New York: W. H. Freeman, 1998. ISBN: 071673026X.

Palmer, E. Laurence, and H. Seymour Fowler. *Fieldbook of Natural History*, 2nd ed. New York: McGraw-Hill, 1975. ISBN: 0070484252.

Parker, Sybil P., ed. *Synopsis and Classification of Living Organisms.* New York: McGraw-Hill, 1982. 2 vols. ISBN: 0070790310.

Tudge, Colin. *The Variety of Life: A Survey and a Celebration of All the Creatures That Have Ever Lived.* Oxford: Oxford University Press, 2000. ISBN: 0198503113.

BIOGRAPHIES

Grinstein, Louise S., et al., eds. *Women in the Biological Sciences: A Biobibliographic Sourcebook.* Westport, CT: Greenwood Press, 1997. ISBN: 0313291802.

Mendelsohn, Everett, ed. *Life Sciences before the Twentieth Century.* New York: Charles Scribner's Sons, 2002. ISBN: 0684806614.

Mendelsohn, Everett, and Brian S. Baigrie, eds. *Life Sciences in the Twentieth Century: Biographical Portraits.* New York: Charles Scribner's Sons, 2001. ISBN: 0684806479.

Shearer, Benjamin F., and Barbara S. Shearer. *Notable Women in the Life Sciences.* Westport, CT: Greenwood Press, 1996. ISBN: 0313293023.

Sterling, Keir B., et al., eds. *Biographical Dictionary of American and Canadian Naturalists and Environmentalists.* Westport, CT: Greenwood Press, 1997. ISBN: 0313230471.

WEB SITES

Botanical Society of America Online Image Collection: http://images.botany.org/

Guild of Natural Science Illustrators: http://www.gnsi.org/

Plants Database, USDA, Natural Resources Conservation Service: http://plants.usda.gov/index.html

University of Georgia's Center for Invasive Species and Ecosystem Health: http://www.insectimages.org/

MISCELLANEOUS

Bobick, James, et al. *The Handy Biology Answer Book.* Detroit: Visible Ink Press, 2004. ISBN: 1578591503.

General Engineering

GUIDE TO THE LITERATURE

Lord, Charles R. *Guide to Information Sources in Engineering.* Englewood, CO: Libraries Unlimited, 2000. ISBN: 1563086999.

HISTORIES

Berlow, Lawrence H. *The Reference Guide to Famous Engineering Landmarks of the World: Bridges, Tunnels, Dams, Roads, and Other Structures.* Phoenix, AZ: Oryx Press, 1998. ISBN: 0897749669.

Davidson, Frank P., and Kathleen Lusk Brooke. *Building the World: An Encyclopedia of the Great Engineering Projects in History.* Westport, CT: Greenwood Press, 2006. ISBN: 0313333734.

Parkyn, Neil, ed. *The Seventy Wonders of the Modern World: 1,500 Years of Extraordinary Feats of Engineering and Construction.* London: Thames & Hudson, 2002. ISBN: 0500510474.

DICTIONARIES

Marchetti, John, ed. *Means Illustrated Construction Dictionary.* Kingston, MA: R. S. Means Co., 2000. ISBN: 0876295383.

From *Science and Technology Resources: A Guide for Information Professionals and Researchers* by James E. Bobick and G. Lynn Berard. Santa Barbara, CA: Libraries Unlimited. Copyright © 2011.

Timings, R. L., and Peter Twigg. *The Pocket Illustrated Dictionary of Engineering Terms.* Oxford: Butterworth Heinemann, 2001. ISBN: 0750650710.

DATABASES

Applied Science and Technology Abstracts (ASTA) (1983 to the present; abstracts 1994 to the present). Applied Science and Technology Abstracts covers core English-language scientific and technical publications. Topics include engineering, acoustics, chemistry, computers, metallurgy, physics, plastics, telecommunications, transportation, and waste management. Periodical coverage includes trade and industrial publications, journals issued by professional and technical societies, and specialized subject periodicals, as well as special issues such as buyers' guides, directories, and conference proceedings.

EI Compendex (also known as the Engineering Index) (1884 to the present). Compendex is the most comprehensive bibliographic database of scientific and technical engineering research available, covering all engineering disciplines. It includes millions of bibliographic citations and abstracts from thousands of engineering journals and conference proceedings. When combined with the Engineering Index Backfile (1884–1969), Compendex covers well over 120 years of core engineering literature.

Environmental Sciences and Pollution Management (1967 to the present).This multidisciplinary database, provides unparalleled and comprehensive coverage of the environmental sciences. Abstracts and citations are drawn from over 10,000 serials including scientific journals, conference proceedings, reports, monographs, books, and government publications.

GeoBase (1980 to the present; Elsevier, Inc.). GeoBase is a multidisciplinary database covering human and physical geography, geology, oceanography, geomechanics, alternative energy sources, pollution, waste management, and nature conservation.

IEEE Xplore (1988 to the present). Provides full-text access to IEEE transactions, journals, and conference proceedings and IEE journals from 1988 to the present, as well as all current IEEE standards. This database provides you with a single source leading to almost one-third of the world's current electrical engineering and computer science literature, with unparalleled access to IEEE and IEE publications.

INSPEC (selected earlier literature to 1896, 1898 to the present). Engineering Information, Inc. Contains literature in electrical engineering, electronics, physics, control engineering, information technology, communications, computers, computing, and manufacturing and production engineering. The database contains nearly 10 million bibliographic records taken from 3,850 scientific and technical journals and 2,200 conference proceedings. Approximately 330,000 new records are added to the database annually.

ISI Web of Science (1945 to the present, coverage varies). Institute for Scientific Information (ISI) http://scientific.thomson.com/isi/. A multidisciplinary database, with searchable author abstracts, covering the literature of the sciences, social sciences, and the arts and humanities. Users may choose to search across all the databases, or select just one. This unique database indexes and links cited references for each article. All formats provide complete bibliographic data and additional features, such as cited reference searching, links to related articles, plus author and publisher addresses.

Knovel: Answers for Science and Engineering (coverage varies). Knovel contains online interactive reference books and databases. It has a database of some of the leading science and engineering reference handbooks, databases, and conference proceedings from publishers such as McGraw-Hill, Elsevier, John Wiley & Sons, ASME, SPE, and ASM International. Copyright: Knovel http://www.knovel.com/

ENCYCLOPEDIAS

McGraw-Hill Concise Encyclopedia of Engineering. New York: McGraw-Hill, 2005. ISBN: 0071439528.

Parker, Sybil P., ed. *McGraw-Hill Encyclopedia of Engineering,* 2nd ed. New York: McGraw-Hill, 1993. ISBN: 0070513929.

HANDBOOKS AND TABLES

Baukal, Charles E., and Robert E. Schwartz, eds. *The John Zink Combustion Handbook.* Boca Raton, FL: CRC Press, 2001. ISBN: 0849323371.

Bird, John O. *Newnes Engineering Science Pocket Book*, 3rd ed. Oxford: Newnes, 2001. ISBN: 0750649917.

Dorf, Richard C., ed. *The Engineering Handbook.* 2nd ed. Boca Raton, FL: CRC Press, 2005. ISBN: 0849315867.

Fire Protection Handbook. Boston, MA: National Fire Protection Association, 1962–. ISSN: 07345984.

Fisk, Arthur, and Wendy A. Rogers, eds. *Handbook of Human Factors and the Older Adult.* San Diego, CA: Academic Press, 1997. ISBN: 0122576802.

Gibilisco, Stan. *Mathematical and Physical Data, Equations, and Rules of Thumb.* New York: McGraw-Hill, 2001. ISBN: 0071361480.

Gieck, Kurt, and Reiner Gieck. *Engineering Formulas*, 7th ed. New York: McGraw-Hill, 1997. ISBN: 007024572X.

Heisler, Sanford I. *The Wiley Project Engineer's Desk Reference: Project Engineering, Operations, and Management.* New York: Wiley, 1994. ISBN: 0471546771.

Kreith, Frank, editor-in-chief. *The CRC Handbook of Thermal Engineering.* Boca Raton, FL: CRC Press, 2000. ISBN: 084939581X.

Levy, Sidney M. *Construction Databook.* New York: McGraw-Hill, 1999. ISBN: 0070383650.

Lewis, Bernard T., and Richard P. Payant. *Facility Inspection Field Manual: A Complete Condition Assessment Guide.* New York: McGraw-Hill, 2001. ISBN: 0071358749.

Lindeburg, Michael R. *Engineering Unit Conversions*, 4th ed. Belmont, CA: Professional Publications, 1999. ISBN: 1888577339.

Parmley, Robert O., ed. *Field Engineer's Manual*, 2nd ed. New York: McGraw-Hill, 1995. ISBN: 0070485798.

Pilkey, Walter D. *Formulas for Stress, Strain, and Structural Matrices.* New York: J. Wiley, 1994. ISBN: 0471527467.

Salvendy, Gavrie. *Handbook of Human Factors and Ergonomics.* New York: J. Wiley, 1997. ISBN: 047116904.

Sims, Frank. *Engineering Formulas: Conversions, Definitions, and Tables.* New York: Industrial Press, 1996. ISBN: 0831130687.

Stanton, Neville, ed. *Handbook of Human Factors and Ergonomics Methods.* Boca Raton, FL: CRC Press, 2005. ISBN: 0415287006.

Thomas, Paul I. *The Contractor's Field Guide*, 2nd ed. Paramus, NJ: Prentice Hall, 2000. ISBN: 0130124168.

Tuma, Jan J., and Ronald A. Walsh. *Engineering Mathematics Handbook*, 4th ed. New York: McGraw-Hill, 1998. ISBN: 0070655294.

Wildi, Théodore. *Units and Conversion Charts: A Handbook for Engineers and Scientists.* New York: IEEE Press, 1991. ISBN: 0879422734.

DIRECTORIES

International Directory of Engineering Societies and Related Organizations, 16th edition. Washington, DC: American Association of Engineering Societies, 1999/2000. ISSN: 10679014.

Trade Shows Worldwide. Detroit, MI: Gale Research Inc., 1990. ISSN: 10464395.

WEB SITES

American Society for Engineering Education: http://www.asee.org

Engineering.com: The Engineer's Ultimate Resource Tool: http://www.engineering.com/Home/tabid/36/Default.aspx

Engineering Subject Gateway: http://vifatec.tib.uni-hannover.de/index.php3?L=e

ICivilEngineer: The Internet for Civil Engineers: http://www.icivilengineer.com/

Intute: Science, Engineering, and Technology: http://www.intute.ac.uk/

National Academy of Engineering: http://www.nae.edu/nae/naehome.nsf

TechXtra: Find articles, key Web sites, books, the latest industry news, job announcements, e-journals, e-prints, technical reports, the latest research, thesis and dissertations: http://www.techxtra.ac.uk/

MISCELLANEOUS

Meenakshi, Raman, and Sangeeta Sharma. *Technical Communication: English Skills for Engineers.* Oxford: Oxford University Press, 2008. ISBN: 0195695747.

Peterson, E. Norman, publisher; Willian D. Machoney, editor-in-chief; L. William Horsley, senior editor. *Means Graphic Construction Standards.* Kingston, MA: R. S.Means Co., 1986. ISBN: 0911950796.

General Science

HISTORY

Schlager, Neil, ed. *Science and Its Times: Understanding the Social Significance of Scientific Discovery.* Detroit, MI: Gale Group, 2000. 8 vols. ISBN: 0787658170.

DICTIONARIES

Clugston, M. J., ed. *The New Penguin Dictionary of Science,* 2nd ed. New York: Penguin Books, 2004. ISBN: 0141010746.

Marvin, Stephen, ed. *Dictionary of Scientific Principles.* San Francisco: Wiley, 2010. ISBN: 978047014680.

McGraw-Hill Dictionary of Scientific and Technical Terms, 6th ed. New York: McGraw-Hill, 2003. ISBN: 007042313X.*Ultimate Visual Dictionary of Science.* New York: DK Publishing, 1998. ISBN: 0789435128.

Walker, Peter M. B., ed. *Larousse Dictionary of Science and Technology.* New York: Larousse, 1995. ISBN: 0752300105.

From *Science and Technology Resources: A Guide for Information Professionals and Researchers* by James E. Bobick and G. Lynn Berard. Santa Barbara, CA: Libraries Unlimited. Copyright © 2011.

ELECTRONIC DATABASES

AccessScience. This is the *McGraw-Hill Encyclopedia of Science and Technology* on the Web. It provides full access to articles, dictionary terms, and hundreds of research updates in all areas of science and technology.

Applied Science and Technology Abstracts (ASTA) (1983 to the present; abstracts 1994 to the present). Applied Science and Technology Abstracts covers core English-language scientific and technical publications. Topics include engineering, acoustics, chemistry, computers, metallurgy, physics, plastics, telecommunications, transportation, and waste management. Periodical coverage includes trade and industrial publications, journals issued by professional and technical societies, and specialized subject periodicals, as well as special issues such as buyers' guides, directories, and conference proceedings.

CSA Illustrata: Natural Sciences (1977 to the present). This Cambridge Scientific Abstracts database indexes tables, figures, graphs, charts, and other illustrations from scholarly literature in agriculture, biology, conservation, earth sciences, environmental studies, fish and fisheries, food and food industries, forests and forestry, geography, medical sciences, meteorology, veterinary science, and water resources since 1997.

Environmental Sciences and Pollution Management (1967 to the present). This multidisciplinary database, provides unparalleled and comprehensive coverage of the environmental sciences. Abstracts and citations are drawn from over 10,000 serials including scientific journals, conference proceedings, reports, monographs, books, and government publications.

GeoBase (1980 to the present; Elsevier, Inc.). GeoBase is a multidisciplinary database covering human and physical geography, geology, oceanography, geomechanics, alternative energy sources, pollution, waste management, and nature conservation.

GeoRef (1966 to the present). Established by the American Geological Institute in 1966, it provides access to the geoscience literature of the world. GeoRef is the most comprehensive database in the geosciences.

HSTM: History of Science, Technology, and Medicine (1975 to the present). Contains references to journal articles, chapters, and reviews within the fields of general science, technology, and medicine. It covers the influence of these fields on society and culture from prehistory to the present.

ISI Web of Science (1945 to the present, coverage varies). Institute for Scientific Information (ISI) http://scientific.thomson.com/isi/. A multidisciplinary database, with searchable author abstracts, covering the literature of the sciences, social sciences, and the arts and humanities. Users may choose to search across all the databases, or select just one. This unique database indexes and links cited references for each article. All formats provide complete bibliographic data and additional features, such as cited reference searching, links to related articles, plus author and publisher addresses.

Medline (1966 to the present). MEDLINE is the National Library of Medicine's (NLM) premier bibliographic database covering the fields of medicine, nursing, dentistry, veterinary medicine, the health care system, and the preclinical sciences. The MEDLINE file contains bibliographic citations and author abstracts from approximately 3,900 current biomedical journals published in the United States and 70 foreign countries.

National Library of the Environment. This is a free, non-advocacy online environmental information service managed by the Committee for the National Institute for the Environment

(CNIE), a non-profit organization (http://www.ncseonline.org/NLE/). CNIE's mission is to improve the scientific basis for making decisions on environmental issues through the successful operation of a National Institute for the Environment (NIE).

NTIS (National Technical Information Services) Indexes. Cambridge Scientific Abstracts (CSA) Indexes reports on government-sponsored R&D from selected federal agencies (e.g., Department of Energy, EPA), their contractors, and their grantees.

Scitopia. Scitopia.org searches more than three and a half million documents, including peer-reviewed journal content and technical conference papers from leading voices in major science and technology disciplines. Searches bibliographic records in each partner's electronic library; patents from the U.S. Patent and Trademark Office, European Patent Office, and Japan Patent Office; and U.S. government documents on the Department of Energy Information Bridge site.

ENCYCLOPEDIAS

Calder, Nigel. *Magic Universe: The Oxford Guide to Modern Science.* Oxford: Oxford University Press, 2003. ISBN: 0198507925.

Considine, Glenn D., ed. *Van Nostrand's Scientific Encyclopedia,* 10th ed. Hoboken, NJ: Wiley, 2008. 3 vols. ISBN: 9780471743385.

Lerner, K. Lee, and Brenda W. Lerner, eds. *The Gale Encyclopedia of Science,* 4th ed. Detroit: Thomson Gale, 2008. 6 vols. ISBN: 9781414428772.

McGraw-Hill Encyclopedia of Science and Technology, 10th ed. New York: McGraw-Hill, 2007. 20 vols. ISBN: 9780071441438.

Meyers, Robert A., ed. *Encyclopedia of Physical Science and Technology*, 3rd ed. San Diego: Academic Press, 2002. 18 vols. ISBN: 0122274105.

Mitcham, Carl, ed. *Encyclopedia of Science, Technology, and Ethics.* Detroit, MI: Macmillan Reference USA, 2005. 4 vols. ISBN: 0028658310.

Nagel, Rob, ed. *U-X-L Encyclopedia of Science,* 2nd ed. Detroit: U-X-L, 2002. 10 vols. ISBN: 0787654329.

Trefil, James S. *The Nature of Science: An A-Z Guide to the Laws and Principles Governing Our Universe.* Boston: Houghton Mifflin, 2003. ISBN: 9780618319381.

HANDBOOKS AND TABLES

Beach, David P. *Handbook for Scientific and Technical Research.* Englewood Cliffs, NJ: Prentice Hall, 1992. ISBN: 0134310403.

Coleman, Gordon J., and David C. Dewar. *The Addison-Wesley Science Handbook.* Reading, MA: Addison-Wesley Publishers, 1997. ISBN: 0201766523.

Day, Robert A., and Barbara Gastel. *How to Write and Publish a Scientific Paper*, 6th ed. Westport, CT: Greenwood Press, 2006. ISBN: 0313330271.

Fisher, David J., ed. *Rules of Thumb for Engineers and Scientists.* Houston, TX: Gulf Pub. Co., 1991. ISBN: 0872017869.

BIOGRAPHIES

American Men and Women of Science, 28th ed. New York: R. R. Bowker, 2010. 8 vols. ISBN: 9781414475519.

Pais, Abraham. *The Genius of Science: A Portrait Gallery.* Oxford: Oxford University Press, 2000. ISBN: 0198506147.

WEB SITES

Academy of Natural Sciences: http://www.ansp.org/

Open Questions: Science News Sites: http://www.openquestions.com/oq-news.htm

PBS: http://www.pbs.org/science/

Popular Science: http://www.popsci.com/

Science Daily: http://www.sciencedaily.com/

Science.gov: http://www.science.gov/

MISCELLANEOUS

Barber, Bernard. *Social Studies of Science.* New Brunswick, NJ: Transaction Publishers, 1990. ISBN: 0887383297.

Bynum, William F., and Roy Porter, eds. *Oxford Dictionary of Scientific Quotations.* Oxford: Oxford University Press, 2005. ISBN: 0198584091.

Fripp, Jon, et al., eds. *Speaking of Science: Notable Quotes on Science, Engineering, and the Environment.* Eagle Rock, VA: LLH Technology Pub., 2000. ISBN: 1878707515.

Gaither, Carl C., and Alma E. Cavazos-Gaither. *Scientifically Speaking: A Dictionary of Quotations.* Philadelphia: Institute of Physics Pub., 2000. ISBN: 075030636X.

Kaplan, Rob, ed. *Science Says: A Collection of Quotations on the History, Meaning, and Practice of Science.* New York: W. H. Freeman, 2001. ISBN: 0716741121.

Krebs, Robert E. *Scientific Laws, Principles, and Theories: A Reference Guide.* Westport, CT: Greenwood Press, 2001. ISBN: 0313309574.

National Academy of Sciences. *Facilitating Interdisciplinary Research.* Washington, DC: The National Academies Press, 2005. ISBN: 0309094356.

National Academy of Sciences. *Integrity in Scientific Research: Creating an Environment that Promotes Responsible Conduct.* Washington, DC: The National Academies Press, 2002. ISBN: 0309085233.

Genetics, Developmental Biology, and Evolutionary Biology

GUIDE TO THE LITERATURE

Young, Christian C., and Mark A. Largent. *Evolution and Creationism: A Documentary and Reference Guide.* Westport, CT: Greenwood Press, 2007. ISBN: 9780313339530.

HISTORIES

Appleman, Philip, ed. *Darwin: Texts Commentary*, 3rd ed. New York: Norton, 2001. ISBN: 0393958493.

Brown, Bryson. *Evolution: A Historical Perspective.* Westport, CT: Greenwood Press, 2007. ISBN: 0313334617.

Everson, Ted. *The Gene: A Historical Perspective.* Westport, CT: Greenwood Press, 2007. ISBN: 0313334498.

Falk, Raphael. *Genetic Analysis: A History of Genetic Thinking.* Cambridge: Cambridge University Press, 2009. ISBN: 9780521884181.

Henig, Robin M. *The Monk in the Garden: The Lost and Found Genuis of Gregor Mendel, the Father of Genetics.* Boston: Houghton Mifflin, 2000. ISBN: 0395977657.

Kay, Lily E. *Who Wrote the Book of Life? A History of the Genetic Code.* Stanford, CA: Stanford University Press, 2000. ISBN: 0804733848.

From *Science and Technology Resources: A Guide for Information Professionals and Researchers* by James E. Bobick and G. Lynn Berard. Santa Barbara, CA: Libraries Unlimited. Copyright © 2011.

Mayr, Ernst. *The Growth of Biological Thought: Diversity, Evolution, and Inheritance.* Cambridge, MA: Belknap Press, 1982. ISBN: 0674364457.

Olby, Robert C. *The Path to the Double Helix: The Discovery of DNA.* New York: Dover Publications, 1994. ISBN: 0486681173.

Peters, James A. *Classic Papers in Genetics.* Englewood Cliffs, NJ: Prentice-Hall, 1959. ISBN: 0131351788.

Portugal, Franklin H., and Jack S. Cohen. *A Century of DNA: A History of the Discovery of the Structure and Function of the Genetic Substance.* Cambridge, MA: MIT Press, 1977. ISBN: 0262160676.

Sturtevant, Alfred H. *A History of Genetics.* Cold Spring Harbor, NY: Cold Spring Harbor Laboratory Press, 2001. ISBN: 0879696079.

Voeller, Bruce R., ed. *The Chromosome Theory of Inheritance: Classic Papers in Development and Heredity.* New York: Appleton-Century-Crofts, 1968. ISBN: 0306500809.

DICTIONARIES

Dye, Frank J. *Dictionary of Developmental Biology and Embryology.* New York: Wiley-Liss, 2002. ISBN: 0471443573.

Farkas, Daniel H. *DNA from A to Z*, 3rd ed. Washington, DC: AACC Press, 2004. ISBN: 1594250022.

King, Robert C., and William D. Stansfield. *A Dictionary of Genetics*, 6th ed. Oxford: Oxford University Press, 2002. ISBN: 0195143248.

Owen, Elizabeth, and Eve Daintith, eds. *The Facts on File Dictionary of Evolutionary Biology.* New York: Facts on File, 2004. ISBN: 0816049246.

Witherly, Jeffre L., et al. *An A to Z of DNA Science: What Scientists Mean When They Talk about Genes and Genomes.* Cold Spring Harbor, NY: Cold Spring Harbor Laboratory Press, 2001. ISBN: 0879696001.

Zhang, Yong-he, and Meng Zhang. *A Dictionary of Gene Technology Terms.* New York: Pantheon, 2001. ISBN: 185070015X.

ELECTRONIC DATABASES

Biosis Previews (1926 to the present). Biosis Previews is a key database for literature searching in biological and life sciences. Covers medical topics with focus on basic research and its coverage of clinical medicine.

Chemical Abstracts (1907 to the present; available electronically via SciFinder Scholar database). SciFinder Scholar is an interface to the *Chemical Abstracts Service* (*CAS*). It allows you to search millions of chemical references and substances by research topic, author name, structure and substructure, chemical name or CAS Registry Number, company/name organization, as well as browse the tables of contents in journals, and enable a reaction query.

ISI Web of Science (1945 to the present, coverage varies). Institute for Scientific Information (ISI) http://scientific.thomson.com/isi/. A multidisciplinary database, with searchable

author abstracts, covering the literature of the sciences, social sciences, and the arts and humanities. Users may choose to search across all the databases, or select just one. This unique database indexes and links cited references for each article. All formats provide complete bibliographic data and additional features, such as cited reference searching, links to related articles, plus author and publisher addresses.

Medline (1966 to the present). MEDLINE is the National Library of Medicine's (NLM) premier bibliographic database covering the fields of medicine, nursing, dentistry, veterinary medicine, the health care system, and the preclinical sciences. The MEDLINE file contains bibliographic citations and author abstracts from approximately 3,900 current biomedical journals published in the United States and 70 foreign countries.

ENCYCLOPEDIAS

Acharya, Tara, and Neeraja Sankaran. *The Human Genome Sourcebook.* Westport, CT: Greenwood Press, 2005. ISBN: 9781573565295.

Brenner, Sydney, and Jeffrey H. Miller, eds. *Encyclopedia of Genetics.* San Diego, CA: Academic Press, 2002. 4 vols. ISBN: 0122270800.

Lerner, K. Lee, and Brenda W. Lerner, eds. *World of Genetics.* Detroit, MI: Gale Group/ Thomson Learning, 2002. 2 vols. ISBN: 0787649589.

Milner, Richard. *Darwin's Universe: Evolution from A to Z.* Berkeley: University of California Press, 2009. ISBN: 9780520243767.

Ness, Bryan D., ed. *Encyclopedia of Genetics.* Rev. ed. Pasadena, CA: Salem Press, 2004. 2 vols. ISBN: 1587651491.

Pagel, Mark D., ed. *Encyclopedia of Evolution.* Oxford, Oxford University Press, 2002. 2 vols. ISBN: 9780195122008.

Regal, Brian, ed. *Icons of Evolution: An Encyclopedia of People, Evidence, and Controversies.* Westport, CT: Greenwood Press, 2008. 2 vols. ISBN: 9780313339110.

Rice, Stanley A. *Encyclopedia of Evolution.* New York: Facts on File, 2007. ISBN: 9780816055159.

Robinson, Richard, ed. *Genetics.* New York: MacMillan Reference USA: Thomson/Gale, 2003. 4 vols. ISBN: 0028656067.

HANDBOOK AND TABLE

Lanza, Robert P., ed. *Handbook of Stem Cells.* Boston: Elsevier Academic, 2004. 2 vols. ISBN: 0124366430.

BIOGRAPHIES

Eldredge, Niles. *Darwin: Discovering the Tree of Life.* New York: W. W. Norton & Co., 2005. ISBN: 0393059669.

Quammen, David. *The Reluctant Mr. Darwin: An Intimate Portrait of Charles Darwin and the Making of His Theory of Evolution.* New York: Norton, 2006. ISBN: 0393059812.

WEB SITES

The Biology Project is an online tutorial for learning biology: http://www.biology.arizona.edu/

Dolan DNA Learning Center: http://www.dnalc.org/

Frequently Asked Questions about Genetic and Genomic Science: http://www.genome.gov/19016904

Genetics Society of America, Web sites of interest: http://www.genetics-gsa.org/pages/sites_of_interest.shtml

National Health Museum: http://www.accessexcellence.org/RC/VL/GG/genes.php

Understanding Genetics, Stanford School of Medicine: http://www.thetech.org/genetics/

MISCELLANEOUS

Brookes, Martin. *Fly: The Unsung Hero of Twentieth-Century Science.* New York: Ecco, 2001. ISBN: 0066212510.

Kohler, Robert E. *Lords of the Fly: Drosophila Genetics and the Experimental Life.* Chicago: University of Chicago Press, 1994. ISBN: 0226450627.

National Academy of Sciences. *Science and Creationism: A View from the National Academy of Sciences,* 2nd ed. Washington, DC: National Academy Press, 1999. ISBN: 0309064066.

General Technology and Inventions

HISTORIES

Dyson, James. *A History of Great Inventions.* New York: Carroll & Graf Publishers, 2001. ISBN: 0786709030.

Evans, Harold. *They Made America: From the Steam Engine to the Search Engine: Two Centuries of Innovators.* Boston: Little, Brown, 2004. ISBN: 0316277665.

Fagan, Brian M., ed. *The Seventy Great Inventions of the Ancient World.* New York: Thames & Hudson, 2004. ISBN: 0500051305.

Harrison, Ian. *The Book of Inventions.* Washington, DC: National Geographic Society, 2004. ISBN: 0792282965.

Philbin, Tom. 1934. *The 100 Greatest Inventions of All Time: A Ranking Past and Present.* New York: Citadel Press, 2003. ISBN: 0806524030.

Stanley, Autumn. *Mothers and Daughters of Invention: Notes for a Revised History of Technology.* Metuchen, NJ: Scarecrow Press, 1993. ISBN: 0810825864.

Van Dulken, Stephen. *American Inventions: A History of Curious, Extraordinary, and Just Plain Useful Patents.* New York: New York University Press, 2004. ISBN: 0814788130.

DATABASES

AccessScience. This is the *McGraw-Hill Encyclopedia of Science and Technology* on the Web. It provides full access to articles, dictionary terms, and hundreds of research updates in all areas of science and technology.Applied Science and Technology Abstracts (ASTA) (1983 to the present; abstracts 1994 to the present). Applied Science and Technology Abstracts covers core English-language scientific and technical publications. Topics include engineering, acoustics, chemistry, computers, metallurgy, physics, plastics, telecommunications, transportation, and waste management. Periodical coverage includes trade and industrial publications, journals issued by professional and technical societies, and specialized subject periodicals, as well as special issues such as buyers' guides, directories, and conference proceedings.

Chemical Abstracts (1907 to the present; available electronically via SciFinder Scholar database). SciFinder Scholar is an interface to the *Chemical Abstracts Service (CAS)*. It allows you to search millions of chemical references and substances by research topic, author name, structure and substructure, chemical name or CAS Registry Number, company/name organization, as well as browse the tables of contents in journals, and enable a reaction query.

EI Compendex (also known as the Engineering Index) (1884 to the present). Compendex is the most comprehensive bibliographic database of scientific and technical engineering research available, covering all engineering disciplines. It includes millions of bibliographic citations and abstracts from thousands of engineering journals and conference proceedings. When combined with the Engineering Index Backfile (1884–1969), Compendex covers well over 120 years of core engineering literature.

Scitopia. Scitopia.org searches more than three and a half million documents, including peer-reviewed journal content and technical conference papers from leading voices in major science and technology disciplines. Searches bibliographic records in each partner's electronic library; patents from the U.S. Patent and Trademark Office, European Patent Office, and Japan Patent Office; and U.S. government documents on the Department of Energy Information Bridge site.U.S. Patent and Trademark Office Patent Databases (1790 to the present). These databases (http://www.uspto.gov/) allow users to search and view the full-text contents and bibliographic information of patents. Users can search by multiple elements of patent records such as patent numbers, titles, inventors, application dates, descriptions/specifications, and so on. The USPTO houses full text for patents issued from 1976 to the present and TIFF images for all patents from 1790 to the present.

ENCYCLOPEDIAS

Carlisle, Rodney P. *Scientific American Inventions and Discoveries: All the Milestones in Ingenuity—From the Discovery of Fire to the Invention of the Microwave Oven.* Hoboken, NJ: J. Wiley, 2004. ISBN: 0471244104.

The Cutting Edge: An Encyclopedia of Advanced Technologies. New York: Oxford University Press, 2000. ISBN: 0195128990.*Encyclopedia of Technology and Applied Sciences.* New York: Marshall Cavendish, 2000. 11 vols. ISBN: 0761471162.

Engelbert, Phillis. *Technology in Action: Science Applied to Everyday Life.* Detroit: U-X-L, 1999. 3 vols. ISBN: 0787628093.

Hempstead, Colin A., ed. *Encyclopedia of 20th-Century Technology.* New York: Routledge, 2005. 2 vols. ISBN: 1579583865.

How Products Are Made: An Illustrated Guide to Product Manufacturing. Detroit: Gale Research, Biennial. ISSN: 10725091.

How Things Work. Geneva, Switzerland: Edito-Service S.A., 1982. 4 vols. ASIN: B000K6VVAW.

Magill, Frank N., ed. *Magill's Survey of Science: Applied Science Series.* Pasadena, CA: Salem Press, 1993. 7 vols. ISBN: 0893567051.

Van Amerongen, C. *The Way Things Work: An Illustrated Encyclopedia of Technology.* New York: Simon & Schuster, 1971. 2 vols. ISBN: 0671210866.

HANDBOOKS AND TABLES

Berinstein, Paula. *The Statistical Handbook on Technology.* Phoenix, AZ: Oryx Press, 1999. ISBN: 1573562084.

Glover, Thomas J. *DeskRef,* 3rd ed. Littleton, CO: Sequoia Pub., 2003. ISBN: 1885071442.

BIOGRAPHIES

Aaseng, Nathan. *Black Inventors.* New York: Facts on File, 1997. ISBN: 0816034079.

Day, Lance, and McNeil, Ian, eds. *Biographical Dictionary of the History of Technology.* New York: Routledge, 1996. ISBN: 0415060427.

Fouché, Rayvon. *Black Inventors in the Age of Segregation: Granville T. Woods, Lewis H. Latimer, & Shelby J. Davidson.* Baltimore: Johns Hopkins University Press, 2003. ISBN: 0801873193.

Hooper, Roger B., ed. *Who's Who of American Inventors.* Baton Rouge, LA: Who's Who of American Inventors/Hooper Group Pub., 1990. ISBN: 0929956044.

Inductees of the National Inventors Hall of Fame, 30th ed. Akron, OH: National Inventors Hall of Fame, 2002. LCCN: 2003-252696.

Sluby, Patricia Carter. *The Inventive Spirit of African Americans: Patented Ingenuity.* Westport, CT: Praeger, 2004. ISBN: 0275966747.

Vare, Ethlie Ann, and Greg Ptacek. *Patently Female: From AZT to TV Dinners: Stories of Women Inventors and Their Breakthrough Ideas.* New York: Wiley, 2002. ISBN: 0471023345.

Zierdt-Warshaw, Linda, et al. *American Women in Technology: An Encyclopedia.* Santa Barbara, CA: ABC-CLIO, 2000. ISBN: 1576070727.

WEB SITES

European Patents: http://ep.espacenet.com/

Free Patents Online: http://www.freepatentsonline.com/

Google Patents: http://www.google.com/patents

Inventor's Handbook: http://web.mit.edu/invent/h-main.html

University of Central Florida University Libraries, Searching Tutorial: http://library.ucf.edu/GovDocs/PatentsTrademarks/Tutorial.asp

U.S. Patent & Trademark Office, Patent Searching Tutorial: http://www.uspto.gov/products/library/ptdl/services/step7.jsp

MISCELLANEOUS

Brown, David E. *Inventing Modern America: From the Microwave to the Mouse.* Cambridge, MA: MIT Press, 2002. ISBN: 0262025086.

How in the World? Pleasantville, NY: Reader's Digest Association, 1990. ISBN: 0895773538.

Langone, John. *The New How Things Work: Everyday Technology Explained.* Washington, DC: National Geographic Society, 2004. ISBN: 079226956X.

Spignesi, Stephen J. *American Firsts: Innovations, Discoveries, and Gadgets Born in the U.S.A.* Franklin Lakes, NJ: New Page Books, 2004. ISBN: 156414691X.

Van Dulken, Stephen. *Inventing the 19th Century: 100 Inventions That Shaped the Victorian Age from Aspirin to the Zeppelin.* New York: New York University Press, 2001. ISBN: 0814788106.

Van Dulken, Stephen. *Inventing the 20th Century: 100 Inventions That Shaped the World: From the Airplane to the Zipper.* Washington Square, NY: New York University Press, 2000. ISBN: 0814788122.

Geology

HISTORIES

Oldroyd, David R. *Earth Cycles: A Historical Perspective.* Westport, CT: Greenwood Press, 2006. ISBN: 0313332290.

Oldroyd, David R. *Sciences of the Earth: Studies in the History of Mineralogy and Geology.* Brookfield, VT: Ashgate, 1998. ISBN: 0860787702.

Thompson, Susan J. *A Chronology of Geological Thinking from Antiquity to 1899.* Metuchen, NJ: Scarecrow Press, 1988. ISBN: 0810821214.

DICTIONARIES

Farndon, John. *Dictionary of the Earth.* New York: Dorling Kindersley, 1994. ISBN: 0751352276.

Glossary of Geology, 4th edition. Alexandria, VA: American Geological Institute, 1997. ISBN: 0922152349.

Manutchehr-Danai, Mohsen. *Dictionary of Gems and Gemology.* New York: Springer, 2000. ISBN: 3540674829.

Watt, Alec. *Longman Illustrated Dictionary of Geology: The Principles of Geology Explained and Illustrated.* Harlow, Essex: Longman, 1982. ISBN: 0582555493.

DATABASES

AAPG Datapages. American Association of Petroleum Geologists. Contains digital databases and utilities for the petroleum geologist. A combined publications database of over 20 geoscience societies. This full-text database covers petroleum geology and related fields.

CSA Illustrata: Natural Sciences (1977 to the present). This Cambridge Scientific Abstracts database indexes tables, figures, graphs, charts, and other illustrations from scholarly literature in agriculture, biology, conservation, earth sciences, environmental studies, fish and fisheries, food and food industries, forests and forestry, geography, medical sciences, meteorology, veterinary science, and water resources since 1997.

EI Compendex (also known as the Engineering Index) (1884 to the present). Compendex is the most comprehensive bibliographic database of scientific and technical engineering research available, covering all engineering disciplines. It includes millions of bibliographic citations and abstracts from thousands of engineering journals and conference proceedings. When combined with the Engineering Index Backfile (1884–1969), Compendex covers well over 120 years of core engineering literature.

GeoBase (1980 to the present; Elsevier, Inc.). GeoBase is a multidisciplinary database covering human and physical geography, geology, oceanography, geomechanics, alternative energy sources, pollution, waste management, and nature conservation.

GeoRef (1966 to the present). Established by the American Geological Institute in 1966, it provides access to the geoscience literature of the world. GeoRef is the most comprehensive database in the geosciences.

ISI Web of Science (1945 to the present, coverage varies). Institute for Scientific Information (ISI) http://scientific.thomson.com/isi/. A multidisciplinary database, with searchable author abstracts, covering the literature of the sciences, social sciences, and the arts and humanities. Users may choose to search across all the databases, or select just one. This unique database indexes and links cited references for each article. All formats provide complete bibliographic data and additional features, such as cited reference searching, links to related articles, plus author and publisher addresses.

Knovel: Answers for Science and Engineering (coverage varies). Knovel contains online interactive reference books and databases. It has a database of some of the leading science and engineering reference handbooks, databases, and conference proceedings from publishers such as McGraw-Hill, Elsevier, John Wiley & Sons, ASME, SPE, and ASM International. Copyright: Knovel http://www.knovel.com/

ENCYCLOPEDIAS

Bowes, Donald R., ed. *The Encyclopedia of Igneous and Metamorphic Petrology.* New York: Van Nostrand Reinhold, 1989. ISBN: 0442206232.

Currie, Philip J., and Kevin Padian, eds. *Encyclopedia of Dinosaurs.* San Diego: Academic Press, 1997. ISBN: 0122268105.

Dasch, E. Julius, ed. *Encyclopedia of Earth Sciences.* New York: Macmillan Reference USA, 1996. 2 vols. ISBN: 0028830008.

Farlow, James O., and M. K. Brett-Surman, eds. *The Complete Dinosaur.* Bloomington: Indiana University Press, 1997. ISBN: 0253333490.

Glut, Donald F. *Dinosaurs, the Encyclopedia.* Jefferson, NC: McFarland & Co., 1997. ISBN: 0899509177.

Good, Gregory A., ed. *Sciences of the Earth: An Encyclopedia of Events, People, and Phenomena.* New York: Garland Pub., 1998. 2 vols. ISBN: 081530062X.

Hancock, Paul L., et al., eds. *Oxford Companion to the Earth.* Oxford: Oxford University Press, 2000. ISBN: 0198540396.

Holden, Martin. *The Encyclopedia of Gemstones and Minerals.* New York: Friedman Group, 1999. ISBN: 1567999492.

Kusky, Timothy M. *Encyclopedia of Earth Science.* New York: Facts on File, 2005. ISBN: 0816049734.

Lambert, David. *Dinosaur Encyclopedia: The Definitive, Fully Illustrated Encyclopedia of Dinosaurs and Other Prehistoric Reptiles.* London: Bloomsbury Books, 1994. ISBN: 1854714503.

Lambert, David. *Encyclopedia of Prehistory.* New York: Facts on File, 2002. ISBN: 081604547X.

Lerner, K. Lee, and Brenda Wilmoth Lerner, eds. *World of Earth Science.* Detroit: Gale/Thomson, 2003. 2 vols. ISBN: 078765681X.

Lessem, Don, and Donald F. Glut. *The Dinosaur Society's Dinosaur Encyclopedia.* New York: Random House, 1993. ISBN: 0679417702.

Middleton, Gerard V. *Encyclopedia of Sediments & Sedimentary Rocks.* Boston: Kluwer Academic Publishers, 2003. ISBN: 1402008724.

Roberts, Willard L., et al. *Encyclopedia of Minerals*, 2nd ed. New York: Van Nostrand Reinhold, 1990. ISBN: 0442276818.

Sigurdsson, Haraldur, ed. *Encyclopedia of Volcanoes.* San Diego: Academic Press, 2000. ISBN: 012643140X.

Singer, Ronald, ed. *Encyclopedia of Paleontology.* Chicago: Fitzroy Dearborn Publishers, 1999. 2 vols. ISBN: 1884964966.

Woodhead, James A. *Earth Science.* Pasadena, CA: Salem Press, 2001. 5 vols. ISBN: 0893560006.

HANDBOOKS AND TABLES

Dana, James D., et al. *Manual of Mineral Science*, 22nd ed. New York: J. Wiley, 2002. ISBN: 0471251771.

Johnsen, Ole. *Minerals of the World.* Princeton, NJ: Princeton University Press, 2002. ISBN: 069109537X.

Schumann, Walter. *Minerals of the World*, 2nd ed. New York: Sterling Pub. Co., 2008. ISBN: 9781402753398.

BIOGRAPHY

Gates, Alexander E. *A to Z of Earth Scientists.* New York: Facts on File, 2002. ISBN: 0816045801.

WEB SITES

Association of American State Geologists: http://www.stategeologists.org/

National Geochemical Database (NGDB): http://minerals.cr.usgs.gov/projects/geochem_data base/

National Geologic Map Database: http://ngmdb.usgs.gov/

OneGeology Portal: http://portal.onegeology.org/

U.S. Board on Geographic Names (BGN): http://geonames.usgs.gov/

MISCELLANEOUS

Barnes-Svarney, Patricia L., and Thomas E. Svarney. *The Handy Geology Answer Book.* Detroit: Visible Ink Press, 2004. ISBN: 1578591562.

History and Philosophy of Science and Technology

GUIDES TO THE LITERATURE

Elliott, Clark A. *History of Science in the United States: A Chronology and Research Guide.* New York: Garland, 1996. ISBN: 0815313098.

Krebs, Robert E. *Scientific Development and Misconception through the Ages: A Reference Guide.* Westport, CT: Greenwood Press, 1999. ISBN: 031330226X.

Olby, Robert C., ed. *Companion to the History of Modern Science.* New York: Routledge, 1990. ISBN: 0415019885.

HISTORIES

Baigrie, Brian S., ed. *History of Modern Science and Mathematics.* New York: Charles Scribner's Sons: Thomson/Gale, 2002. 4 vols. ISBN: 0684806363.

Bunch, Bryan H., and Alexander Hellemans. *The History of Science and Technology: A Browser's Guide to the Great Discoveries, Inventions, and the People Who Made Them, From the Dawn of Time to Today.* Boston: Houghton Mifflin, 2004. ISBN: 0618221239.

Engelbert, Phillis. *Science, Technology, and Society: The Impact of Science in the 20th Century.* Detroit: UXL/Thomson/Gale, 2002. 3 vols. ISBN: 0787656496.

Knight, Judson, and Neil Schlager, eds. *Science, Technology, and Society: The Impact of Science from 2000 B.C. to the 18th Century.* Detroit: UXL/Thomson/Gale, 2002. 4 vols. ISBN: 0787656534.

Krebs, Robert E. *Groundbreaking Scientific Experiments, Inventions, and Discoveries of the Middle Ages and the Renaissance.* Westport, CT: Greenwood Press, 2004. ISBN: 0313324336.

Krebs, Robert E., and Carolyn A. Krebs. *Groundbreaking Scientific Experiments, Inventions, and Discoveries of the Ancient World.* Westport, CT: Greenwood Press, 2003. ISBN: 0313313423.

Magill, Frank N., ed. *Great Events from History II: Science and Technology.* Pasadena, CA: Salem Press, 1991. 5 vols. ISBN: 0893566373.

Newton, David E., et al., eds. *Science, Technology, and Society: The Impact of Science in the 19th Century.* Detroit: U-X-L, 2001. 2 vols. ISBN: 0787648744.

Shectman, Jonathan. *Groundbreaking Scientific Experiments, Inventions, and Discoveries of the 18th Century.* Westport, CT: Greenwood Press, 2003. ISBN: 0313320152.

Suplee, Curt. *Milestones of Science.* Washington, DC: National Geographic Society, 2000. ISBN: 0792279069.

Windelspecht, Michael. *Groundbreaking Scientific Experiments, Inventions, and Discoveries of the 17th Century.* Westport, CT: Greenwood Press, 2002. ISBN: 0313315019.

Windelspecht, Michael. *Groundbreaking Scientific Experiments, Inventions, and Discoveries of the 19th Century.* Westport, CT: Greenwood Press, 2003. ISBN: 0313319693.

BIBLIOGRAPHIES

Black, George W. *American Science and Technology: A Bicentennial Bibliography.* Carbondale: Southern Illinois University, 1979. ISBN: 0809308983.

Rothenberg, Marc. *The History of Science and Technology in the United States: A Critical and Selective Bibliography, Volume 1.* New York: Garland Pub., 1982. ISBN: 0824092783.

Rothenberg, Marc. *The History of Science and Technology in the United States: A Critical and Selective Bibliography, Volume 2.* New York: Garland Pub., 1993. ISBN: 0824083490.

DICTIONARY

Sebastian, Anton. *A Dictionary of the History of Science.* New York: Parthenon Group, 2001. ISBN: 185070418X.

ELECTRONIC DATABASES

Biosis Previews (1926 to the present). Biosis Previews is a key database for literature searching in biological and life sciences. Covers medical topics with focus on basic research and its coverage of clinical medicine.

HSTM: History of Science, Technology, and Medicine (1975 to the present). Contains references to journal articles, chapters, and reviews within the fields of general science, technology, and medicine. It covers the influence of these fields on society and culture from prehistory to the present.

ISI Web of Science (1945 to the present, coverage varies). Institute for Scientific Information (ISI) http://scientific.thomson.com/isi/. A multidisciplinary database, with searchable author abstracts, covering the literature of the sciences, social sciences, and the arts and humanities. Users may choose to search across all the databases, or select just one. This unique database indexes and links cited references for each article. All formats provide complete bibliographic data and additional features, such as cited reference searching, links to related articles, plus author and publisher addresses.

Medline (1966 to the present). MEDLINE is the National Library of Medicine's (NLM) premier bibliographic database covering the fields of medicine, nursing, dentistry, veterinary medicine, the health care system, and the preclinical sciences. The MEDLINE file contains bibliographic citations and author abstracts from approximately 3,900 current biomedical journals published in the United States and 70 foreign countries.

ENCYCLOPEDIAS

Applebaum, Wilbur, ed. *Encyclopedia of the Scientific Revolution: From Copernicus to Newton.* New York: Garland Pub., 2000. ISBN: 0815315031.

Burns, William E. *The Scientific Revolution: An Encyclopedia.* Santa Barbara, CA: ABC-CLIO, 2001. ISBN: 0874368758.

Heilbron, J. L., ed. *The Oxford Companion to the History of Modern Science.* Oxford: Oxford University Press, 2003. ISBN: 0195112296.

Hessenbruch, Arne, ed. *Reader's Guide to the History of Science.* Chicago: Fitzroy Dearborn, 2000. ISBN: 188496429X.

McNeil, Ian, ed. *An Encyclopaedia of the History of Technology.* New York: Routledge, 1990. ISBN: 0415013062.

Rothenberg, Marc, ed. *The History of Science in the United States: An Encyclopedia.* New York: Garland, 2001. ISBN: 0815307624.

HANDBOOKS AND TABLES

Asimov, Isaac. *Asimov's Chronology of Science & Discovery.* New York: HarperCollins, 1994. ISBN: 0062701134.

Clark, John, et al. *Philip's Science & Technology: People, Dates & Events.* London: George Philip Ltd., 1999. ISBN: 054007716X.

Hellemans, Alexander, and Bryan Bunch. *The Timetables of Science: A Chronology of the Most Important People and Events in the History of Science.* New, updated. New York: Simon & Schuster, 1991. ISBN: 0671733281.

BIOGRAPHIES

Allaby, Michael, and Derek Gjertsen. *Makers of Science.* New York: Oxford University Press, 2002. 5 vols. ISBN: 0195216806.

American Council of Learned Societies. *Concise Dictionary of Scientific Biography*, 2nd ed. New York: Scribner's, 2000. ISBN: 0684806312.

Bersanelli, Marco, and Mario Gargantini. *From Galileo to Gell-Mann: The Wonder That Inspired the Greatest Scientists of All Time in Their Own Words.* West Conshohocken, PA: Templeton Press, 2009. ISBN: 1599473402.

Day, Lance, and Ian McNeil, eds. *Biographical Dictionary of the History of Technology.* New York: Routledge, 1996. ISBN: 0415060427.

WEB SITES

Echo is a major Web site that catalogues, annotates, and reviews sites on the history of science, technology, and medicine: http://echo.gmu.edu/

Further Resources in History of Medicine, from the National Library of Medicine: http://www.nlm.nih.gov/hmd/resources/index.html

History of the Health Sciences, Web sites provided by the Medical Library Association (MLA): http://mla-hhss.org/histlink.htm

MISCELLANEOUS

Bud, Robert, et al., eds. *Instruments of Science: An Historical Encyclopedia.* New York: Science Museum and National Museum of American History, in association with Garland Pub., 1998. ISBN: 0815315619.

Downs, Robert B. *Landmarks in Science: Hippocrates to Carson.* Littleton, CO: Libraries Unlimited, 1982. ISBN: 0872879255.

Frasca-Spada, Marina, and Nick Jardine, eds. *Books and the Sciences in History.* Cambridge: Cambridge University Press, 2000. ISBN: 0521650631.

Industrial Engineering

GUIDE TO THE LITERATURE

Finley, Elsie, comp.; Marcia Parsons, jt. comp. *Guide to Literature on Industrial Engineering.* Washington: American Society for Engineering Education, Engineering School Libraries Division, 1970.

DICTIONARIES

McGraw-Hill Dictionary of Engineering, 2nd ed. [electronic resource]. New York: McGraw-Hill, 2003. ISBN: 0071417990.

Philippsborn, H. E. *Elsevier's Dictionary of Industrial Technology.* In English, German, and Portuguese. New York: Elsevier, 1994. ISBN: 0444899456.

Polon, David D., ed. *Dictionary of Industrial Engineering Abbreviations, Signs and Symbols.* New York: Odyssey Press, [1967]. LCCN: 65-19183.

DATABASES

ABI Inform (1970 to the present; ProQuest). This database contains articles from scholarly and trade sources and dissertations on business conditions and trends, management practice and theory, corporate strategy and tactics, and the competitive landscape.

Applied Science and Technology Abstracts (ASTA) (1983 to the present; abstracts 1994 to the present). Applied Science and Technology Abstracts covers core English-language scientific and technical publications. Topics include engineering, acoustics, chemistry, computers, metallurgy, physics, plastics, telecommunications, transportation, and waste management. Periodical coverage includes trade and industrial publications, journals issued by professional and technical societies, and specialized subject periodicals, as well as special issues such as buyers' guides, directories, and conference proceedings.

EI Compendex (also known as the Engineering Index) (1884 to the present). Compendex is the most comprehensive bibliographic database of scientific and technical engineering research available, covering all engineering disciplines. It includes millions of bibliographic citations and abstracts from thousands of engineering journals and conference proceedings. When combined with the Engineering Index Backfile (1884–1969), Compendex covers well over 120 years of core engineering literature.

IEEE Xplore (1988 to the present). Provides full-text access to IEEE transactions, journals, and conference proceedings and IEE journals from 1988 to the present, as well as all current IEEE standards. This database provides you with a single source leading to almost one-third of the world's current electrical engineering and computer science literature, with unparalleled access to IEEE and IEE publications.

INSPEC (selected earlier literature to 1896, 1898 to the present). Engineering Information, Inc. Contains literature in electrical engineering, electronics, physics, control engineering, information technology, communications, computers, computing, and manufacturing and production engineering. The database contains nearly 10 million bibliographic records taken from 3,850 scientific and technical journals and 2,200 conference proceedings. Approximately 330,000 new records are added to the database annually.

ISI Web of Science (1945 to the present, coverage varies). Institute for Scientific Information (ISI) http://scientific.thomson.com/isi/. A multidisciplinary database, with searchable author abstracts, covering the literature of the sciences, social sciences, and the arts and humanities. Users may choose to search across all the databases, or select just one. This unique database indexes and links cited references for each article. All formats provide complete bibliographic data and additional features, such as cited reference searching, links to related articles, plus author and publisher addresses.

Materials Research Database with Metadex (1966 to the present; Cambridge Scientific Abstracts). This database includes leading materials science databases with specialist content on materials science, metallurgy, ceramics, polymers, and composites used in engineering applications. The collection provides coverage on applied and theoretical materials processes including welding and joining, heat treatment, and thermal spray. Everything from raw materials and refining through processing, welding, and fabrication to end-use, corrosion, performance, and recycling is covered in depth.

ENCYCLOPEDIAS

Cochran, James J. *Wiley Encyclopedia of Operations Research and Management Science* [Wiley InterScience: electronic resource]. Hoboken, NJ: J. Wiley, 2010.

Industrial Engineering Applications & Practice Users' Encyclopedia. International Journal of Industrial Engineering, 1998. ISBN: 0965450600.

McGraw-Hill Encyclopedia of Environmental Science & Engineering, 3rd ed. New York: McGraw-Hill, c1993. ISBN: 0070513961.

HANDBOOKS AND TABLES

Eshbach, Ovid W. *Eshbach's Handbook of Engineering Fundamentals,* 5th ed. [Knovel: electronic resource]. Hoboken, NJ: John Wiley, c2009. ISBN: 1601198396.

Hicks, Tyler Gregory. *Standard Handbook of Engineering Calculations,* 4th ed. New York: McGraw-Hill, c2005. ISBN: 0071427937.

Mansdorf, S. Z. *Complete Manual of Industrial Safety.* Englewood Cliffs, NJ: Prentice Hall, 1993. ISBN: 0131596330.

Salvendy, Gavriel, ed. *Handbook of Industrial Engineering: Technology and Operations Management,* 3rd ed. New York: Wiley, 2001. ISBN: 0471330574.

Zandin, Kjell B., ed. *Maynard's Industrial Engineering Handbook,* 5th ed. New York: McGraw-Hill, 2001. ISBN: 0070411026.

WEB SITES

American Productivity and Quality Center (APQC): http://www.apqc.org/

American Society for Quality, Quality Resource Directory: http://asq.org/perl/vqn/vqn_search.cgi

Corporate Information: http://www.corporateinformation.com/Country-Industry-Research-Links.aspx

Informs Online, Institute for Operations Research and Management Science: http://www.informs.org/

Institute of Industrial Engineers: http://www.iienet2.org/Default.aspx

The Society of Manufacturing Engineers: http://www.sme.org/cgi-bin/getsmepg.pl?/new-sme.html&&&SME&

Mathematics

GUIDES TO THE LITERATURE

Fowler, Kristine K. *Using the Mathematics Literature.* New York: Marcel Dekker, 2004. ISBN: 0824750357.

Tucker, Martha A., and Nancy D. Anderson. *Guide to Information Sources in Mathematics and Statistics.* Santa Barbara, CA: Libraries Unlimited, 2004. ISBN: 1563087014.

HISTORIES

Anglin, W. S. *Mathematics: A Concise History and Philosophy,* 2nd ed. New York: Springer-Verlag, 1996. ISBN: 0387942807.

Boyer, Carl B. *A History of Mathematics,* 2nd ed. Revised by Uta C. Merzbach. Hoboken, NJ: John Wiley & Sons, 1991. ISBN: 0471543977.

Bruno, Leonard C., and Lawrence W. Baker. *Math & Mathematicians: The History of Math Discoveries around the World.* Detroit: U.X.L., 1999. 2 vols. ISBN: 0787638129.

Cooke, Roger. *The History of Mathematics: A Brief Course,* 2nd ed. Hoboken, NJ: John Wiley& Sons, 2005. ISBN: 0471444596.

Grattan-Guinness, I., ed. *Companion Encyclopedia of the History and Philosophy of the Mathematical Sciences.* New York: Routledge, 1994. 2 vols. ISBN: 0415037859.

Grattan-Guinness, I. *The Norton History of the Mathematical Sciences: The Rainbow of Mathematics.* New York: W. W. Norton, 1998. ISBN: 0393046508.

Katz, Victor J. *A History of Mathematics: An Introduction,* 2nd ed. River, NJ:Addison-Wesley, 1998. ISBN: 0321016181.

Mankiewicz, Richard. *The Story of Mathematics*, new ed. Princeton: Princeton University Press, 2004. ISBN: 0691120463.

Stillwell, John. *Mathematics and Its History,* 2nd ed. New York: Springer, 2004. ISBN: 0387953361.

BIBLIOGRAPHIES

Dauben, Joseph W. *The History of Mathematics from Antiquity to the Present: A Selective Bibliography.* New York: Garland, 1985. ISBN: 0824092848.

Gaffney, Matthew P., and Lynn Arthur Steen. *Annotated Bibliography of Expository Writing in the Mathematical Sciences.* Washington, DC: Mathematical Association of America, 1978. ISBN: 0883854228.

Karpinski, Louis Charles. *Bibliography of Mathematical Works Printed in America through 1850.* Ann Arbor: University of Michigan Press, 1940. National Library: 012149431.

Schaaf, William L. *A Bibliography of Recreational Mathematics,* 4th ed. Reston, VA: National Council of Teachers of Mathematics,1970. ISBN: 0873530209.

Smith, David Eugene, ed. *Rara Arithmetic: A Catalogue of the Arithmetics Written before the Year MDCI [1601], with a Description of Those in the Library of George Arthur Plimpton of New York,* 4thed. New York: Cosimo Classics, 2007. ISBN: 1602066906.

Steen, Lynn Arthur. *Library Recommendations for Undergraduate Mathematics.* Washington, DC: Mathematical Association of America, 1992. ISBN: 0883850761.

Steen, Lynn Arthur. *Mathematics Books Recommendations for High School and Public Libraries.*Washington, DC: Mathematical Association of America, 1992. ISBN: 0883854554.

Thiessen, Diane, and Margaret Matthias. *The Wonderful World of Mathematics: A Critically Annotated List of Children's Books in Mathematics.* Reston, VA: National Council of Teachers of Mathematics, 1992. ISBN: 0873533534.

DICTIONARIES

Berry, John. *Dictionary of Mathematics.* New York: Fitzroy Dearborn Publishers, 1999. ISBN: 1579581579.

Daintith, John, and John Clark. *The Facts on File Dictionary of Mathematics,* 4th ed. [print and electronic resource]. New York: Facts on File, 1999. ISBN: 081605651X.

James, Robert C. *Mathematics Dictionary,* 5th ed. New York: Van Nostrand Reinhold, 1992. ISBN: 0442007418.

Krantz, Steven G. *Dictionary of Algebra, Arithmetic, and Trigonometry.* Boca Raton, FL: CRC Press, 2001. ISBN: 158488052X.

Mathematical Society of Japan. *Encyclopedic Dictionary of Mathematics,* 2nd ed. Boston: MIT Press, 1993. 2 vols. ISBN: 0262590204.

Nelson, David. *The Penguin Dictionary of Mathematic,* 3rd ed. New York: Penguin Books, 2003. ISBN: 0141010770.

ELECTRONIC DATABASES

ISI Web of Science (1945 to the present, coverage varies). Institute for Scientific Information (ISI) http://scientific.thomson.com/isi/. A multidisciplinary database, with searchable author abstracts, covering the literature of the sciences, social sciences, and the arts and humanities. Users may choose to search across all the databases, or select just one. This unique database indexes and links cited references for each article. All formats provide complete bibliographic data and additional features, such as cited reference searching, links to related articles, plus author and publisher addresses.

Knovel: Answers for Science and Engineering (coverage varies). Knovel contains online interactive reference books and databases. It has a database of some of the leading science and engineering reference handbooks, databases, and conference proceedings from publishers such as McGraw-Hill, Elsevier, John Wiley & Sons, ASME, SPE, and ASM International. Copyright: Knovel http://www.knovel.com/

MATHnetBASE. Searchable full-text editions of over 150 CRC/Chapman & Hall texts, handbooks, dictionaries, and tables covering mathematics and engineering mathematics. Copyright: CRC Press/Chapman Hall www.crcpress.com/

MathSciNet (1940 to the present; American Mathematical Society). It **is** an electronic publication offering access to a carefully maintained and easily searchable database of reviews, abstracts, and bibliographic information for much of the mathematical sciences literature. Over 80,000 new items are added each year, most of them classified according to the Mathematics Subject Classification, and contains over 2 million items and over 700,000 direct links to original articles. It continues in the tradition of the paper publication *Mathematical Reviews* (MR), which was first published in 1940. MathSciNet bibliographic data from retro-digitized articles dates back to 1864.

ENCYCLOPEDIAS

Brandenberger, Barry Max. *Mathematics.* Farmington Hills, MI: Macmillan Reference USA: Gale Group/Thomson Learning, 2002. 4 vols. ISBN: 0028655613.

Hazewinkel, Michiel, ed. *Encyclopaedia of Mathematics.* New York: Springer, 1995. 12 vols. and index vol. ISBN:-13: 978-0792334989.

Narins, Brigham. *World of Mathematics.* Farmington Hills, MI: Gale Group, 2001. 2 vols. ISBN: 0787636525.

Tanton, James Stuart. *Encyclopedia of Mathematics* [print and electronic resource]. New York: Facts on File, 2005. ISBN: 0816051240.

Weisstein, Eric W. *CRC Concise Encyclopedia of Mathematics,* 2nd ed. Boca Raton, FL: Chapman & Hall/CRC, 2003. ISBN: 1584883472.

HANDBOOKS AND TABLES

Beyer, William H. *CRC Handbook of Mathematical Sciences,* 6th ed. Boca Raton, FL: CRC Press, 1987. ISBN: 0849306566.

Pearson, Carl E. *Handbook of Applied Mathematics: Selected Results and Methods,* 2nd, rev. ed. New York: Van Nostrand Reinhold, Co., 1990. ISBN: 0442005210.

Råde, Lennart. *Mathematics Handbook for Science and Engineering,* 5th ed. [print and electronic resource]. Berlin: Springer; Studentlitteratur, 2004. ISBN: 3540211411.

Zwillinger, Daniel, editor-in-chief. *CRC Standard Mathematical Tables and Formulae,* 31st ed. Boca Raton, FL: CRC Press, 2003. ISBN: 1584882913.

BIOGRAPHIES

Albers, Donald J., and G. L. Alexanderson. *Mathematical People: Profiles and Interviews,* 2nd ed. Switzerland: Birkhauser, 2008. ISBN: 9781568813400.

Albers, Donald J., et al. *More Mathematical People: Contemporary Conversations.* San Diego: Academic Press, 1994. ISBN: 0120482517.

Franceschetti, Donald R. *Biographical Encyclopedia of Mathematicians.* Tarrytown, NY:Marshall Cavendish, 1999. 2 vols. ISBN: 0761470697.

Henrion, Claudia. *Women in Mathematics: The Addition of Difference.* Bloomington: Indiana University Press, 1997. ISBN: 0253211190.

Hollingdale, S. H. *Makers of Mathematics.* New York: Penguin Books, 1994. ISBN: 0140149228.

Morrow, Charlene, and Teri Perl, eds. *Notable Women in Mathematics: A Biographical Dictionary.* Westport, CT: Greenwood Press, 1998. ISBN: 0313291314.

Young, Robyn V. *Notable Mathematicians: From Ancient Times to the Present.* Farmington Hills, MI: Gale, 1998. ISBN: 0787630713.

DIRECTORIES

American Mathematical Society. *Mathematical Sciences Professional Directory.* Annual. ISSN: 07374356.

International Mathematical Union, ed. *World Directory of Mathematicians.* American Mathematical Society, 2002. ISBN: 9992067039.

Jaguszewski, Janice M. *Recognizing Excellence in the Mathematical Sciences: An International Compilation of Awards, Prizes, and Recipients.* Stamford, CT: JAI Press, 1997. ISBN: 0762302356.

WEB SITES

AMS Digital Mathematics Registry: http://www.ams.org/dmr/. The aim of the AMS-DMR is to provide centralized access to certain collections of digitized publications in the mathematical sciences. The registry is primarily focused on older material from journals and journal-like book series that originally appeared in print but are now available in digital form. The registry is organized both by the collections and by the individual journals (or series) themselves, providing links to each that will be regularly verified and updated.

The Math Forum Internet Mathematics Library@Drexel University: http://mathforum.org/ library/. The Math Forum is the leading online resource for improving math learning, teaching, and communication since 1992. The contributors are teachers, mathematicians, researchers, students, and parents using the power of the Web to learn math and improve math education.

Wolfram MathWorld, the Web's most extensive mathematics resource: http://mathworld. wolfram.com/. MathWorld continues to be the most popular and most visited mathematics site on the Internet and its mathematical content continues to steadily grow and expand. *MathWorld*™ is the Web's most extensive mathematical resource, provided as a free service to the world's mathematics and Internet communities as part of a commitment to education and educational outreach by Wolfram Research, makers of *Mathematica.*

Zentralblatt MATH: http://www.emis.de/ (via the EMS homepage). This is the world's most complete and longest-running abstracting and reviewing service in pure and applied mathematics. The Zentralblatt MATH database contains more than 2 million entries drawn from more than 2,300 serials and journals and covers the period from 1868 to the present by the recent integration of the Jahrbuch database (JFM). The entries are classified according to the Mathematics Subject Classification Scheme (MSC 2000).

MISCELLANEOUS

Barrow, John D. *100 Essential Things You Didn't Know You Didn't Know: Math Explains Your World.* New York: W. W. Norton & Co., 2009. ISBN: 0393070077.

Bellos, Alex. *Here's Looking at Euclid: A Surprising Excursion through the Astonishing World of Math.* New York: Free Press, 2010. ISBN: 9781416588252.

Bentley, Peter J. *The Book of Numbers: The Secret of Numbers and How They Changed the World.* Toronto: Firefly Books, 2008. ISBN: 1554073618.

Blatner, David. *The Joy of [pi].* New York: Walker and Co., 1999. ISBN: 0802775624.

Dorrie, Heinrich, and David Antin. *100 Great Problems of Elementary Mathematics: Their History and Solutions.* New York: Dover Publications, 1965. ISBN: 0486613488.

Dudley, Underwood. *Mathematical Cranks.* Cambridge: Cambridge University Press, 1996. ISBN: 0883855070.

Gardner, Martin. *The Colossal Book of Mathematics: Classic Puzzles, Paradoxes, and Problems: Number Theory, Algebra, Geometry, Probability, Topology, Game Theory, Infinity, and Other Topics of Recreational Mathematics.* New York: Norton, 2001. ISBN: 0393020231.

Gibilisco, Stan. *Everyday Math Demystified.* Columbus, OH: McGraw-Hill, 2004. ISBN: 0071431195.

Hartston, William. *The Book of Numbers: The Ultimate Compendium of Facts about Figures*, 2nd ed. New York: MetroBooks, 2000. ISBN: 1900512939.

Hopkins, Nigel J., John W.Mayne, and John R.Hudson. *Go Figure! The Numbers You Need for Everyday Life.* Canton, MI: Visible Ink Press, 1992. ISBN: 0810394243.

Nowlan, Robert A. *A Dictionary of Quotations in Mathematics.* Jefferson, NC: McFarland, 2002. ISBN: 0786412844.

Paulos, John Allen. *Innumeracy: Mathematical Illiteracy and Its Consequences.* New York: Hill and Wang, 2001. ISBN: 0809058405.

Pickover, Clifford A. *The Math Book: From Pythagoras to the 57th Dimension, 250 Milestones in the History of Mathematics.* New York: Sterling, 2009. ISBN: 1402757964.

Salem, Lionel, et al. *The Most Beautiful Mathematical Formulas: An Entertaining: ook at the Most Insightful, Useful and Quirky Theorems of All Time.* Hoboken, NJ: John Wiley& Sons, 1997. ISBN: 0471176621.

Singh, Simon. *Fermat's Enigma: The Epic Quest to Solve the World's Greatest Mathematical Problem.* New York: Anchor Books, 1998. ISBN: 0385493622.

Stewart, Ian. *Professor Stewart's Cabinet of Mathematical Curiosities.* New York: Basic Books, 2008. ISBN: 0465013023.

Stewart, Ian. *Professor Stewart's Hoard of Mathematical Treasures.* New York: Basic Books, 2010. ISBN: 0465017754.

Verma, Surendra. *The Little Book of Maths Theorems, Theories & Things.* Australia: New Holland Publishing, 2009. ISBN: 1741106710.

Wilson, Robin J. *Stamping through Mathematics.* New York: Springer, 2001. ISBN: 0387989498.

Materials Science Engineering

DICTIONARIES

Davis, Joseph R., ed. *ASM Materials Engineering Dictionary.* Metals Park, OH: ASM International, 1992. ISBN: 0871704471.

Keller, Harald, and Uwe Erb. *Dictionary of Engineering Materials.* Hoboken, NJ: Wiley-Interscience, 2004. ISBN: 0471444367.

DATABASES

Chemical Abstracts (1907 to the present; available electronically via SciFinder Scholar database). SciFinder Scholar is an interface to the *Chemical Abstracts Service* (*CAS*). It allows you to search millions of chemical references and substances by research topic, author name, structure and substructure, chemical name or CAS Registry Number, company/name organization, as well as browse the tables of contents in journals, and enable a reaction query.EI Compendex (also known as the Engineering Index) (1884 to the present). Compendex is the most comprehensive bibliographic database of scientific and technical engineering research available, covering all engineering disciplines. It includes millions of bibliographic citations and abstracts from thousands of engineering journals and conference proceedings. When combined with the Engineering Index Backfile (1884–1969), Compendex covers well over 120 years of core engineering literature.

ISI Web of Science (1945 to the present, coverage varies). Institute for Scientific Information (ISI) http://scientific.thomson.com/isi/. A multidisciplinary database, with searchable

author abstracts, covering the literature of the sciences, social sciences, and the arts and humanities. Users may choose to search across all the databases, or select just one. This unique database indexes and links cited references for each article. All formats provide complete bibliographic data and additional features, such as cited reference searching, links to related articles, plus author and publisher addresses.

Knovel: Answers for Science and Engineering (coverage varies). Knovel contains online interactive reference books and databases. It has a database of some of the leading science and engineering reference handbooks, databases, and conference proceedings from publishers such as McGraw-Hill, Elsevier, John Wiley & Sons, ASME, SPE, and ASM International. Copyright: Knovel http://www.knovel.com/

Materials Research Database with Metadex (1966 to the present; Cambridge Scientific Abstracts). This database includes leading materials science databases with specialist content on materials science, metallurgy, ceramics, polymers, and composites used in engineering applications. The collection provides coverage on applied and theoretical materials processes including welding and joining, heat treatment, and thermal spray. Everything from raw materials and refining through processing, welding, and fabrication to end-use, corrosion, performance, and recycling is covered in depth.

NTIS (National Technical Information Services) Indexes. Cambridge Scientific Abstracts (CSA) Indexes reports on government-sponsored R&D from selected federal agencies (e.g., Department of Energy, EPA), their contractors, and their grantees.

Scitopia. Scitopia.org searches more than three and a half million documents, including peer-reviewed journal content and technical conference papers from leading voices in major science and technology disciplines. Searches bibliographic records in each partner's electronic library; patents from the U.S. Patent and Trademark Office, European Patent Office, and Japan Patent Office; and U.S. government documents on the Department of Energy Information Bridge site.U.S. Patent and Trademark Office Patent Databases (1790 to the present). These databases (http://www.uspto.gov/) allow users to search and view the full-text contents and bibliographic information of patents. Users can search by multiple elements of patent records such as patent numbers, titles, inventors, application dates, descriptions/specifications, and so on. The USPTO houses full text for patents issued from 1976 to the present and TIFF images for all patents from 1790 to the present.

ENCYCLOPEDIAS

Cahn, R. W, ed. *Concise Encyclopedia of Materials Charaterization.* New York: Elsevier Science, 2005. ISBN: 0080445470.

Cahn, R. W, and David Bloor, eds. *Encyclopedia of Advanced Materials.* New York: Pergamon Press, 1994. ISBN: 0080406068.

Cahn, R. W., et al., eds. *Materials Science and Technology: A Comprehensive Treatment.* Weinheim, Germany: VCH, 1997. 18 vols. ISBN: 3527268138.

HANDBOOKS AND TABLES

Ash, Michael, and Irene Ash. *Handbook of Corrosion Inhibitors.* Endicott, NY: Synapse Information Resources, 2001. ISBN: 1890595241.

ASM International Handbook Committee. *ASM Handbook,* 10th ed. Materials Park, OH: ASM International, 1990. 22 vols. ISBN: 0871703777.

ASM International Handbook Committee. *Engineered Materials Handbook.* Metals Park, OH: ASM International, 1997. 4 vols. ISSN: 10602372.

ASM International Materials Properties Database Committee. *ASM Ready Reference: Properties and Units for Engineering Alloys.* Materials Park, OH: ASM International, 1997. ISBN: 0871705850.

Bauccio, Michael, edr. *ASM Engineering Materials Reference Book,* 2nd ed. Materials Park, OH: ASM International, 1994. ISBN: 0871705028.

Bauccio, Michael, ed. *ASM Metals Reference Book,* 3rd ed. Materials Park, OH: ASM International, 1993. ISBN: 0871704781.

Brady, George S., and Henry R. Clauser. *Materials Handbook: An Encyclopedia for Purchasing Agents, Engineers, Executives, and Foremen,* 14th ed. New York: McGraw-Hill, 1991.

Buch, Alfred. *Pure Metals Properties: A Scientific-Technical Handbook.* Materials Park, OH: ASM International, 1999. ISBN: 0871706377.

Cardarelli, François. *Materials Handbook: A Concise Desktop Reference.* London: Springer, 2000. ISBN: 1852331682.

Davis, Joseph R., ed. *Alloy Digest Sourcebook: Stainless Steels.* Materials Park, OH: ASM International, 2000. ISBN: 0871706490.

Davis, Joseph R., ed. *Concise Metals Engineering Data Book.* Materials Park, OH: ASM International, 1997. ISBN: 0871706067.

Davis, Joseph R., ed. *Copper and Copper Alloys.* Materials Park, OH: ASM International, 2001. ISBN: 0871707268.

Davis, Joseph R., ed. *Heat-resistant Materials.* Materials Park, OH: ASM International, 1997. ISBN: 0871705966.

Davis, Joseph R., ed. *Metals Handbook,* 2nd ed. Materials Park, OH: ASM International, 1998. ISBN: 0871706547.

Davis, Joseph R., ed. *Stainless Steels.* Materials Park, OH: ASM International, 1994. ISBN: 0871705036.

Frick, John P., ed. *Woldman's Engineering Alloys,* 9th ed. Materials Park, OH: ASM International, 2000. ISBN: 0871706911.

Gauthier, Michelle M. *Engineered Materials Handbook.* Materials Park, OH: ASM International, 1995. ISBN: 0871702835.

Harvey, Philip D., ed. *Engineering Properties of Steel.* Metals Park, OH: American Society for Metals, 1982.

Kreysa, G., and M. Schütze, eds. *Corrosion Handbook: Corrosive Agents and Their Interaction with Materials,* 2nd ed. Weinheim, Germany: Wiley-VCH, 2004. 13 vols. ISBN: 3527311173.

Mallick, P. K., ed. *Composites Engineering Handbook.* New York: M. Dekker, 1997. ISBN: 0824793048.

Nayar, Alok. *The Metals Databook.* New York: McGraw-Hill, 1997. ISBN: 0070460884.

Revie, R. Winston, and Herbert Henry Uhlig, eds. *Uhlig's Corrosion Handbook,* 2nd ed. New York: Wiley, 2000. ISBN: 0471157775.

Ross, Robert B. *Metallic Materials Specification Handbook*, 4th ed. London: Chapman & Hall, 1992. ISBN: 0412369400.

Shackelford, James F., and William Alexander, eds. *CRC Materials Science and Engineering Handbook,* 3rd ed. Boca Raton, FL: CRC Press, 2001. ISBN: 0849326966.

Walsh, Ronald A. *McGraw-Hill Machining and Metalworking Handbook.* New York: McGraw-Hill, 1994. ISBN: 0070679584.

WEB SITES

Galaxy.com: http://www.galaxy.com/dir29736/Materials_Science.htm

International Union of Crystallography: http://ww1.iucr.org/cww-top/crystal.index.html

MatWeb, Material Property Data: http://www.matweb.com/

National Institute of Standards and Technology: http://www.nist.gov/index.html

MISCELLANEOUS

Annual Book of ASTM Standards. This series from ASTM (the American Society for Testing and Materials) gives detailed testing methods for a host of properties of different types of materials and objects. Volumes are grouped by the substance/object tested. There is a subject index to locate individual standards, as well as a numeric index for locating the volume of a particular standard number.

Mechanical Engineering

BIBLIOGRAPHIES

Franklin, Hugh Lockwood. *Selective Guide to Literature on Mechanical Engineering.* American Society for Engineering Education, Engineering Libraries Division, c1985. ISBN: 0878231021.

DICTIONARIES

Homer, J. G., and G. K. Grahame-White. *A Dictionary of Mechanical Engineering Terms*, 9th ed. London: Technical P., 1967. ISBN: 0291393578.

Nayler, G. H. F. *Dictionary of Mechanical Engineering*, 4th ed. Warrendale, PA: Society of Automotive Engineers, 1996. ISBN: 1560917547.

Polon, David D. *Dictionary of Mechanical Engineering Abbreviations, Signs and Symbols.* New York: Odyssey Press, 1967. ASIN: B002BOHHSQ.

DATABASES

Applied Mechanics Reviews (1948 to the present, updated monthly). American Society of Mechanical Engineers. Reviews of books and journal articles related to mechanical engineering. Includes complete review articles.

EI Compendex (also known as the Engineering Index) (1884 to the present). Compendex is the most comprehensive bibliographic database of scientific and technical engineering research available, covering all engineering disciplines. It includes millions of bibliographic citations and abstracts from thousands of engineering journals and conference proceedings. When combined with the Engineering Index Backfile (1884–1969), Compendex covers well over 120 years of core engineering literature.

IEEE Xplore (1988 to the present). Provides full-text access to IEEE transactions, journals, and conference proceedings and IEE journals from 1988 to the present, as well as all current IEEE standards. This database provides you with a single source leading to almost one-third of the world's current electrical engineering and computer science literature, with unparalleled access to IEEE and IEE publications.

INSPEC (selected earlier literature to 1896, 1898 to the present). Engineering Information, Inc. Contains literature in electrical engineering, electronics, physics, control engineering, information technology, communications, computers, computing, and manufacturing and production engineering. The database contains nearly 10 million bibliographic records taken from 3,850 scientific and technical journals and 2,200 conference proceedings. Approximately 330,000 new records are added to the database annually.

ISI Web of Science (1945 to the present, coverage varies). Institute for Scientific Information (ISI) http://scientific.thomson.com/isi/. A multidisciplinary database, with searchable author abstracts, covering the literature of the sciences, social sciences, and the arts and humanities. Users may choose to search across all the databases, or select just one. This unique database indexes and links cited references for each article. All formats provide complete bibliographic data and additional features, such as cited reference searching, links to related articles, plus author and publisher addresses.

Knovel: Answers for Science and Engineering (coverage varies). Knovel contains online interactive reference books and databases. It has a database of some of the leading science and engineering reference handbooks, databases, and conference proceedings from publishers such as McGraw-Hill, Elsevier, John Wiley & Sons, ASME, SPE, and ASM International. Copyright: Knovel http://www.knovel.com/

NTIS (1964 to the present, updated monthly). The National Technical Information Service database is the preeminent, official resource for government-sponsored U.S. and worldwide scientific, technical, engineering, and business-related information.

SAE Publications and Standards Database (1906 to the present). This database contains bibliographic data for more than 69,000 documents published by the Society of Automotive Engineers, including technical papers since 1906, magazine articles, books, all ground vehicle and aerospace standards, specifications, and research reports.

ENCYCLOPEDIAS

Davidson, Frank Paul. *Building the World: An Encyclopedia of the Great Engineering Projects in History.* Westport, CT: Greenwood Press, 2006. ISBN: 0313333734 (vol. 1); 0313333742 (vol. 2).

Webster, John G. *Encyclopedia of Medical Devices & Instrumentation.* Hoboken, NJ: Wiley InterScience, c2006. ISBN: 0471676004.

Wnek, Gary E., and Gary L. Bowlin, eds. *Encyclopedia of Biomaterials and Biomedical Engineering.* New York: Marcel Dekker, c2004. ISBN: 0824754980 (vol. 1); 0824755561 (vol. 2).

HANDBOOKS AND TABLES

Amiss, John M., et al. *Guide to the Use of Tables and Formulas in Machinery's Handbook*, 26th ed. New York: Industrial Press, 2000. ISBN: 083112699X.

Carvill, James. *Mechanical Engineer's Data Handbook.* Boca Raton, FL: CRC Press, 1993. ISBN: 0849377803.

Frankel, Michael. *Facility Piping Systems Handbook,* 2nd ed. New York: McGraw-Hill, 2002. ISBN: 0071358773.

Hicks, Tyler G., ed. *Handbook of Mechanical Engineering Calculations.* New York: McGraw-Hill, 1998. ISBN: 0070288135.

Hoogers, Gregor, ed. *Fuel Cell Technology Handbook.* Boca Raton, FL: CRC Press, 2003. ISBN: 0849308771.

Karassik, Igor J., et al., eds. *Pump Handbook*, 3rd ed. New York: McGraw-Hill, 2001. ISBN: 0070340323.

Marks, Lionel S. *Marks' Standard Handbook for Mechanical Engineers*, 11th ed. New York: McGraw-Hill, 2007. ISBN: 9780071428675.

Matthews, Clifford. *ASME Engineer's Data Book.* New York: ASME Press, 2001. ISBN: 0791801551.

Nayyar, Mohinder L., ed. *Piping Handbook,* 7th ed. New York: McGraw-Hill, 2000. ISBN: 0070471061.

Rothbart, Harold A., ed. *Mechanical Design Handbook.* New York: McGraw-Hill, 1996. ISBN: 0070540381.

Stein, Benjamin, and John S. Reynolds. *Mechanical and Electrical Equipment for Buildings*, 9th ed. New York: Wiley, 2000. ISBN: 0471156965.

Timings, Roger L. *Newnes Workshop Engineer's Pocket Book.* Oxford: Newnes, 2000. ISBN: 0750647191.

Timings, Roger L., and Tony May. *Newnes Mechanical Engineer's Pocket Book*, 2nd ed. Oxford: Newnes, 1997. ISBN: 0750632623.

Woodson, R. Dodge. *Plumber's Field Manual.* New York: McGraw-Hill, 1997. ISBN: 0070717796.

Woodson, R. Dodge. *Plumber's Standard Handbook.* New York: McGraw-Hill, 1999. ISBN: 0071343865.

WEB SITES

Fatigue Calculator: https://efatigue.com/

Mechanical Engineering Design Resources Web site: http://www.gearhob.com/

Reliability Engineering Resource Web site: http://www.weibull.com/

Scitopia.org: http://www.scitopia.org/scitopia/

MISCELLANEOUS

Xianguo, Li. *Principles of Fuel Cells.* New York: Taylor & Francis, 2006. ISBN: 1591690226.

Meteorology

DICTIONARY

Glickman, Todd S. *Glossary of Meteorology,* 2nd ed. Boston: American Meteorological Society, 2000. ISBN: 1878220349.

DATABASES

AMS Journal Online. American Meteorological Society. This database is a freely searchable site that contains the full text of all AMS journals (fee-based access to the full text via subscription) (http://journals.ametsoc.org/). The entire journal database—including accepted manuscripts not yet formally published—is searchable by all, with access restricted to subscribers for the full-text Early Online Release manuscripts and articles from the most recent two years.

ISI Web of Science (1945to the present, coverage varies). Institute for Scientific Information (ISI) http://scientific.thomson.com/isi/. A multidisciplinary database, with searchable author abstracts, covering the literature of the sciences, social sciences, and the arts and humanities. Users may choose to search across all the databases, or select just one. This unique database indexes and links cited references for each article. All formats provide complete bibliographic data and additional features, such as cited reference searching, links to related articles, plus author and publisher addresses.

Oceanic Abstracts (1981 to the present; Cambridge Scientific Abstracts). The database focuses on marine biology and physical oceanography, fisheries, and aquaculture. This database is totally comprehensive in its coverage of living and nonliving resources, and meteorology and geology, plus environmental, technological, and legislative topics.

ENCYCLOPEDIAS

Allaby, Michael. *Encyclopedia of Weather and Climate.* New York: Facts on File, 2007. 2 vols. ISBN: 0816063508.

Engelbert, Phillis. *The Complete Weather Resource.* Detroit: UXL, 1997. 4 vols. ISBN: 0810397870.

Fry, Juliane L., et al. *The Encyclopedia of Weather and Climate Change: A Complete Visual Guide.* Berkeley, CA: University of California Press, 2010. ISBN: 9780520261013.

Newton, David E. *Encyclopedia of Air.* Westport, CT: Greenwood Press, 2003. ISBN: 1573565644.

Philander, George S., ed. *Encyclopedia of Global Warming and Climate Change.* Los Angeles: SAGE, 2008. 3 vols. ISBN: 9781412958783.

Schneider, Stephen H., ed. *Encyclopedia of Climate and Weather.* New York: Oxford University Press, 1996. 2 vols. ISBN: 0195094859.

HANDBOOKS AND TABLES

Garoogian, David. *Weather America: A Thirty-Year Summary of Statistical Weather Data and Rankings,* 2nd ed. Lakeville, CT: Grey House Pub., 2001. ISBN: 1891482297.

Pearce, E.A., and C.G. Smith. *Fodor's World Weather Guide.* New York: Random House, 1999. ISBN: 0375703535.

Ruffner, James A., and Frank E. Bair. *The Weather Almanac,* 10th ed. Detroit: Gale Research Co., 2001. ISSN: 07315627.

BIOGRAPHY

Rittner, Don. *A to Z of Scientists in Weather and Climate.* New York: Facts on File, 2003. ISBN: 0816047979.

WEB SITES

Astrophysics Data Service, Smithsonian Astrophysical Observatory (SAO): http://adswww. harvard.edu/

National Oceanic and Atmospheric Administration, National Weather Service: http://www. nws.noaa.gov/

NOAA Central Library, National Oceanographic Data Center: http://www.lib.noaa.gov/researchtools/subjectguides/wind/windandsea.html

WW2010 (Weather World 2010 Project): http://ww2010.atmos.uiuc.edu/%28Gh%29/home.rxml

MISCELLANEOUS

Hile, Kevin. *The Handy Weather Answer Book,* 2nd ed. Canton, MI: Visible Ink Press, 2009. ISBN: 9781578592210.

Microbiology

HISTORY

Brock, Thomas D., ed. *Milestones in Microbiology, 1546–1940.* Washington, DC: ASM Press, 1999. ISBN: 1555811426.

DICTIONARIES

Herbert, John W., et al., eds. *Dictionary of Immunology*, 4th edn. San Diego: Academic Press, 1995. ISBN: 0127520252.

Singleton, Paul, and Diana Sainsbury. *Dictionary of Microbiology and Molecular Biology*, 2nd ed. New York: Wiley, 1993. ISBN: 0471940526.

ELECTRONIC DATABASES

Algology, Mycology, and Protozoology Abstracts (1982 to the present; Cambridge Scientific Abstracts). The scope of coverage includes reproduction, growth, life cycles, biochemistry, genetics, and infection and immunity in humans, other animals, and plants.

Bacteriology Abstracts (1982 to the present; ProQuest). Covering topics ranging from bacterial immunology and vaccinations to diseases of humans and animals, the journal provides access to far-reaching clinical findings as well as all aspects of pure bacteriology, biochemistry, and genetics. Main coverage: Bacteriology and microbiology, bacterial immunology, vaccinations, antibiotics, and immunology.

Biosis Previews (1926 to the present). Biosis Previews is a key database for literature searching in biological and life sciences. Covers medical topics with focus on basic research and its coverage of clinical medicine.

CSA Illustrata: Natural Sciences (1977 to the present). This Cambridge Scientific Abstracts database indexes tables, figures, graphs, charts, and other illustrations from scholarly literature in agriculture, biology, conservation, earth sciences, environmental studies, fish and fisheries, food and food industries, forests and forestry, geography, medical sciences, meteorology, veterinary science, and water resources since 1997.

ISI Web of Science (1945 to the present, coverage varies). Institute for Scientific Information (ISI) http://scientific.thomson.com/isi/. A multidisciplinary database, with searchable author abstracts, covering the literature of the sciences, social sciences, and the arts and humanities. Users may choose to search across all the databases, or select just one. This unique database indexes and links cited references for each article. All formats provide complete bibliographic data and additional features, such as cited reference searching, links to related articles, plus author and publisher addresses.

Medline (1966 to the present). MEDLINE is the National Library of Medicine's (NLM) premier bibliographic database covering the fields of medicine, nursing, dentistry, veterinary medicine, the health care system, and the preclinical sciences. The MEDLINE file contains bibliographic citations and author abstracts from approximately 3,900 current biomedical journals published in the United States and 70 foreign countries.

ENCYCLOPEDIAS

Delves, Peter J., and Ivan M. Roitt, eds. *Encyclopedia of Immunology*, 2nd ed. San Diego: Academic Press, 1998. 4 vols. ISBN: 0122267656.

Granoff, Allan, and Robert G. Webster, eds. *Encyclopedia of Virology*, 2nd ed. San Diego: Academic Press, 1999. 3 vols. ISBN: 0122270304.

Lederberg, Joshua, ed. *Encyclopedia of Microbiology*, 2nd ed. San Diego: Academic Press, 2000. 4 vols. ISBN: 0122268008.

Lerner, K. Lee, and Brenda W. Lerner, eds. *World of Microbiology and Immunology.* Detroit: Gale, 2003. 2 vols. ISBN: 0787665401.

Robinson, Richard K., et al., eds. *Encyclopedia of Food Microbiology.* San Diego: Academic Press, 2000. 3 vols. ISBN: 0122270703.

HANDBOOKS AND TABLES

Atlas, Ronald M., and Lawrence C. Parks. *Handbook of Microbiological Media.* Boca Raton, FL: CRC Press, 1993. ISBN: 0849329442.

Garrity, George M., et al., eds. *Bergey's Manual of Systematic Bacteriology*, 2nd ed. New York: Springer, 2001. 3 vols. ISBN: 0387897711.

WEB SITES

Biomed Experts: http://www.biomedexperts.com/Portal.aspx

The NCBI Entrez Taxonomy homepage: http://www.ncbi.nlm.nih.gov/sites/entrez?db=tax onomy

VADLO is brought to you by Life in Research, LLC, a company founded by two biology scientists who wish to make it easier to locate biology research–related information on the Web: http://vadlo.com/

MISCELLANEOUS

Sankaran, Neeraja. *Microbes and People: An A-Z of Microorganisms in Our Lives.* Phoenix, AZ: Oryx Press, 2000. ISBN: 1573562173.

Nutrition and Food Science

HISTORIES

Benning, Lee E. *The Cook's Tales: Origins of Famous Foods and Recipes.* Old Saybrook, CT: Globe Pequot Press, 1992. ISBN: 0871062291.

Flandrin, Jean-Louis, and Massimo Montanari. *Food: A Culinary History from Antiquity to the Present.* New York: Columbia University Press, 1999. ISBN: 0231111541.

Kiple, Kenneth F., and Kriemhild C. Ornelas, eds. *The Cambridge World History of Food.* Cambridge: Cambridge University Press, 2000. 2 vols. ISBN: 0521402166.

Stevens, Patricia B. *Rare Bits: Unusual Origins of Popular Recipes.* Athens: Ohio University Press, 1998. ISBN: 0821412329.

Ward, Susie, et al. *The Gourmet Atlas: The History, Origin, and Migration of Foods of the World.* New York: Macmillan, 1997. ISBN: 0028619889.

DICTIONARIES

Bender, David A. *Bender's Dictionary of Nutrition and Food Technology,* 8th ed. Boca Raton, FL: CRC Press, 2006. ISBN: 0849376017.

Drummond, Karen E. *Dictionary of Nutrition and Dietetics.* New York: Van Nostrand Reinhold, 1996. ISBN: 0442022255.

From *Science and Technology Resources: A Guide for Information Professionals and Researchers* by James E. Bobick and G. Lynn Berard. Santa Barbara, CA: Libraries Unlimited. Copyright © 2011.

Lagua, Rosalinda T., and Virginia S. Claudio. *Nutrition and Diet Therapy Reference Dictionary,* 4th ed. New York: Chapman & Hall, 1996. ISBN: 0412070510.

Mariani, John F. *The Dictionary of American Food and Drink.* New York: Hearst Books, 1994. ISBN: 0688099963.

Palmatier, Robert A. *Food: A Dictionary of Literal and Nonliteral Terms.* Westport, CT: Greenwood Press, 2000. ISBN: 0313314365.

Rolland, Jacques. *The Cook's Essential Kitchen Dictionary: A Complete Culinary Resource.* Toronto: R. Rose, 2004. ISBN: 0778800989.

DATABASES

Biosis Previews (1926 to the present). Biosis Previews is a key database for literature searching in biological and life sciences. Covers medical topics with focus on basic research and its coverage of clinical medicine.

CINAHL®: The Cumulative Index to Nursing and Allied Health Literature (1981 to the present, EBSCO). This database is the most comprehensive resource for nursing and allied health literature. *CINAHL* subject headings follow the structure of the Medical Subject Headings, or MeSH, used by the National Library of Medicine.

ISI Web of Science (1945 to the present, coverage varies). Institute for Scientific Information (ISI) http://scientific.thomson.com/isi/. A multidisciplinary database, with searchable author abstracts, covering the literature of the sciences, social sciences, and the arts and humanities. Users may choose to search across all the databases, or select just one. This unique database indexes and links cited references for each article. All formats provide complete bibliographic data and additional features, such as cited reference searching, links to related articles, plus author and publisher addresses.

Medline (1966 to the present). MEDLINE is the National Library of Medicine's (NLM) premier bibliographic database covering the fields of medicine, nursing, dentistry, veterinary medicine, the health care system, and the preclinical sciences. The MEDLINE file contains bibliographic citations and author abstracts from approximately 3,900 current biomedical journals published in the United States and 70 foreign countries.

Nutrition and Food Sciences Database. CABI, Centre for Agriculture and Biosciences International. This is a specialist database covering human nutrition, food science, and food technology. No other database can provide such a comprehensive view of the food chain or of the interactions between diet and health. This database contains more than 765,000 records dating back to 1973, with over 50,000 added annually. It annually selects records from over 5,000 serials, as well as 400–500 nonserial publications, covering literature from 125 countries. It also covers books, conference proceedings, bulletins, reports, and published theses.

ENCYCLOPEDIAS

Bijlefeld, Marjolijn, and Sharon K. Zoumbaris. *Encyclopedia of Diet Fads.* Westport, CT: Greenwood Press, 2003. ISBN: 0313322236.

Cassell, Dana K., and David H. Gleaves. *The Encyclopedia of Obesity and Eating Disorders*, 3rd ed. New York: Facts on File, 2006. ISBN: 0816061971.

Considine, Douglas M., and Glenn D. Considine, eds. *Foods and Food Production Encyclopedia.* New York: Van Norstrand Reinhold, 1982. ISBN: 0442216122.

Davidson, Alan. *The Penguin Companion to Food.* Rev. ed. New York: Penguin Books, 2002. ISBN: 0140515224.

Davidson, Alan, and Tom Jaine, eds. *The Oxford Companion to Food*, 2nd ed. Oxford: Oxford University Press, 2006. ISBN: 0192806815.

Duyff, Roberta L. *American Dietetic Association Complete Food and Nutrition Guide,* 3rd ed. Hoboken, NJ: John Wiley & Sons, 2006. ISBN: 0470041153.

Fortin, François, ed. *The Visual Food Encyclopedia.* New York: Macmillan, 1996. ISBN: 0028610067.

Gilman, Sander L. *Diets and Dieting: A Cultural Encyclopedia.* New York: Routledge, 2008. ISBN: 0415974208.

James, Delores C. S., ed. *Nutrition and Well-Being A to Z.* Detroit, MI: Macmillan Reference USA, 2004. 2 vols. ISBN: 0028657071.

Keller, Kathleen, ed. *Encyclopedia of Obesity.* Los Angeles: Sage Publications, 2008. 2 vols. ISBN: 1412952387.

Larousse Gastronomique: The World's Greatest Culinary Encyclopedia. New York: Clarkson Potter Publishers, 2009. ISBN: 0307464911.

Longe, Jacqueline L., ed. *The Gale Encyclopedia of Diets: A Guide to Health and Nutrition.* Detroit: Thomson/Gale, 2008. 2 vols. ISBN: 1414429916.

Navarra, Tova. *The Encyclopedia of Vitamins, Minerals, and Supplements,* 2nd ed. New York: Facts on File, Inc., 2004. ISBN: 081604998X.

PDR for Nutritional Supplements, 2nd ed. Montvale, NJ: Medical Economics, Thomson Healthcare, 2008. ISSN: 15343642.

Petit, William A., Jr., and Christine Adamec. *The Encyclopedia of Diabetes.* New York: Facts on File, 2002. ISBN: 0816044988.

Rinzler, Carol A. *The New Complete Book of Food: A Nutritional, Medical, and Culinary Guide,* 2nd ed. New York: Facts on File, 2009. ISBN: 081607710X.

Ronzio, Robert A. *The Encyclopedia of Nutrition and Good Health,* 2nd ed. New York: Facts on File, 2003. ISBN: 0816049661.

Russell, Percy J., and Anita Williams. *The Nutrition and Health Dictionary.* New York: Chapman & Hall, 1995. ISBN: 0412989816.

Sadler, Michele, ed. *Encyclopedia of Human Nutrition.* San Diego: Academic Press, 1999. 3 vols. ISBN: 0122266943.

Smith, Andrew F. *Encyclopedia of Junk Food and Fast Food.* Westport, CT: Greenwood Press, 2006. ISBN: 0313335273.

Smith, Andrew F., ed. *The Oxford Encyclopedia of Food and Drink in America.* Oxford: Oxford University Press, 2004. 2 vols. ISBN: 0195154371.

Snodgrass, Mary E. *Encyclopedia of Kitchen History.* New York: Fitzroy Dearborn, 2004. ISBN: 1579583806.

HANDBOOKS AND TABLES

Ash, Michael, and Irene Ash. *Handbook of Food Additives*, 2nd ed. Endicott, NY: Synapse Information Resources, 2002. ISBN: 1890595365.

Bowes, Anna De Planter, and Charles F. Church. *Bowes and Church's Food Values of Portions Commonly Used*, 19th ed. Philadelphia: Lippincott, 2010. ISBN: 9780781781343.

Holzmeister, Lea Ann. *The Diabetes Carbohydrate & Fat Gram Guide: Quick, Easy Meal Planning Using Carbohydrate & Fat Gram Counts,* 4th ed. Alexandria, VA: American Diabetes Association, 2010. ISBN: 1580403409.

Raghavan, Susheela. *Handbook of Spices, Seasonings, & Flavorings,* 2nd ed. New York: CRC Press, 2006. ISBN: 084932842X.

Rinzler, Carol A. *The New Complete Book of Herbs, Spices, and Condiments.* New York: Facts on File, 2001. ISBN: 0816041539.

Roberts, Cynthia A. *The Food Safety Information Handbook.* Westport, CT: Greenwood Press, 2001. ISBN: 1573563056.

BIOGRAPHY

Arndt, Alice, ed. *Culinary Biographies: A Dictionary of the World's Great Historic Chefs, Cookbook Authors and Collectors, Farmers, Gourmets, Home Economists, Nutritionists, Restaurateurs, Philosophers, Physicians, Scientists, Writers, and Others Who Influenced the Way We Eat Today.* Houston: Yes Press, 2006. ISBN: 0971832218.

WEB SITES

Center for Disease Control and Prevention, Division of Nutrition, Physical Activity and Obesity: http://www.cdc.gov/nccdphp/dnpao/index.html

HealthFinder, U.S. Department of Health & Human Services: http://www.healthfinder.gov/

National Agricultural Library, Food and Nutrition: http://riley.nal.usda.gov/nal_display/index.php?info_center=8&tax_level=1&tax_subject=2

National Library of Medicine: http://www.nlm.nih.gov/

U.S. Food and Drug Administration, Center for Food Safety and Applied Nutrition: http://www.fda.gov/Food/default.htm

MISCELLANEOUS

Barnette, Martha. *Ladyfingers & Nun's Tummies: A Lighthearted Look at How Foods Got Their Names.* New York: Times Books, 1997. ISBN: 0812921003.

Livingston, A. D., and Helen Livingston. *Edible Plants and Animals: Unusual Foods from Aardvark to Zamia.* New York: Facts on File, 1993. ISBN: 0816027447.

Mahan, L. Kathleen, and Sylvia Escott-Stump. *Krause's Food, Nutrition, & Diet Therapy,* 11th ed. Philadelphia: W. B. Saunders, 2004. ISBN: 0721697844.

Matthews, Dawn D. *Diabetes Sourcebook*, 3rd ed. Detroit, MI: Omnigraphics, 2003. ISBN: 0780806298.

Rodriquez, Judith C. *The Diet Selector: How to Choose a Diet Perfectly Tailored to Your Needs.* Philadelphia: Running Press, 2007. ISBN: 0762431709.

Shannon, Joyce B. *Diet and Nutrition Sourcebook*, 3rd ed. Detroit, MI: Omnigraphics, 2006. ISBN: 0780808002.

Physics

GUIDE TO THE LITERATURE

Stern, David. *Guide to Information Sources in the Physical Sciences.* Englewood, CO: Libraries Unlimited, 2000. ISBN: 1563087510.

HISTORIES

Atkins, Stephen E. *Historical Encyclopedia of Atomic Energy.* Westport, CT: Greenwood Press, 2000. ISBN: 0313304009.

Baigrie, Brian. *Electricity and Magnetism: A Historical Perspective.* Westport, CT: Greenwood Press, 2006. ISBN: 0313333580.

Lewis, Christopher J. T. *Heat and Thermodynamics: A Historical Perspective.* Westport, CT: Greenwood Press, 2007. ISBN: 0313333327.

Peacock, Kent A. *The Quantum Revolution: A Historical Perspective.* Westport, CT: Greenwood Press, 2007. ISBN: 031333448X.

Shore, Steven N. *Forces in Physics: A Historical Perspective.* Westport, CT: Greenwood Press, 2008. ISBN: 0313333033.

From *Science and Technology Resources: A Guide for Information Professionals and Researchers* by James E. Bobick and G. Lynn Berard. Santa Barbara, CA: Libraries Unlimited. Copyright © 2011.

DICTIONARIES

Challoner, Jack. *The Visual Dictionary of Physics.* New York: Dorling Kindersley, 1995. ISBN: 0789402394.

Cullerne, John, et al., eds. *The Penguin Dictionary of Physics,* 3rd edition. New York: Penguin, 2000. ISBN: 0140514597.

Darton, Mike, and John Clark. *The Macmillan Dictionary of Measurement.* New York: Maxwell Macmillan International, 1994. ISBN: 0025257501.

Fenna, Donald. *A Dictionary of Weights, Measures, and Units.* New York: Oxford University Press, 2002. ISBN: 0198605226.

McGraw-Hill Dictionary of Physics, 3rd edition. New York: McGraw-Hill, 2003. ISBN: 0071410481.

Rennie, Richard, editor. *The Facts on File Dictionary of Atomic and Nuclear Physics.* New York: Facts on File, 2003. ISBN: 0816049165.

DATABASES

arXiv.org (originally developed by Paul Ginsparg and started in 1991). Provides open access to 626,887 e-prints in physics, mathematics, computer science, quantitative biology, and statistics. arXiv, Open Archives Iniative (OAI) compliant, is owned, operated, and funded by Cornell University, and is partially funded by the National Science Foundation.

Chemical Abstracts (1907 to the present; available electronically via SciFinder Scholar database). SciFinder Scholar is an interface to the *Chemical Abstracts Service (CAS).* It allows you to search millions of chemical references and substances by research topic, author name, structure and substructure, chemical name or CAS Registry Number, company/name organization, as well as browse the tables of contents in journals, and enable a reaction query.

INSPEC (selected earlier literature to 1896, 1898 to the present). Engineering Information, Inc. Contains literature in electrical engineering, electronics, physics, control engineering, information technology, communications, computers, computing, and manufacturing and production engineering. The database contains nearly 10 million bibliographic records taken from 3,850 scientific and technical journals and 2,200 conference proceedings. Approximately 330,000 new records are added to the database annually.

ISI Web of Science (1945 to the present, coverage varies). Institute for Scientific Information (ISI) http://scientific.thomson.com/isi/. A multidisciplinary database, with searchable author abstracts, covering the literature of the sciences, social sciences, and the arts and humanities. Users may choose to search across all the databases, or select just one. This unique database indexes and links cited references for each article. All formats provide complete bibliographic data and additional features, such as cited reference searching, links to related articles, plus author and publisher addresses.

Knovel: Answers for Science and Engineering (coverage varies). Knovel contains online interactive reference books and databases. It has a database of some of the leading science and engineering reference handbooks, databases, and conference proceedings from publishers such as McGraw-Hill, Elsevier, John Wiley & Sons, ASME, SPE, and ASM International. Copyright: Knovel http://www.knovel.com/

MathSciNet (1940 to the present; American Mathematical Society). A searchable database providing access to over 55 years of *Mathematical Reviews* and *Current Mathematical Publications*. MathSciNet contains over 2 million items and over 700,000 direct links to original articles in the mathematical sciences.

ENCYCLOPEDIAS

Besançon, Robert M., ed. *The Encyclopedia of Physics,* 3rd ed. New York: Van Nostrand Reinhold Co., 1995. ISBN: 0442257783.

Carlisle, Rodney P., ed. *Encyclopedia of the Atomic Age.* New York: Facts on File, 2001. ISBN: 081604029X.

Lerner, Rita G., and George L. Trigg, eds. *Encyclopedia of Physics,* 2nd ed. New York: VCH, 1991. ISBN: 0895737523.

Lord, John. *Sizes: The Illustrated Encyclopedia.* New York: HarperPerennial, 1995. ISBN: 0062732285.

Rigden, John S., ed. *Macmillan Encyclopedia of Physics.* New York: Simon & Schuster Macmillan, 1996. 4 vols. ISBN: 0028657039.

Rosen, Joe. *Encyclopedia of Physics.* New York: Facts on File, 2004. ISBN: 0816049742.

World of Physics. Detroit: Gale Group, 2001. 2 vols. ISSN: 15310809.

HANDBOOKS AND TABLES

Alenitsyn, Alexander G., et al. *Concise Handbook of Mathematics and Physics.* Moscow: Nauka Publishers, 1997. ISBN: 0849377455.

Benenson, Walter, et al., eds. *Handbook of Physics.* New York: Springer, 2002. ISBN: 0387952691.

Grigoriev, Igor S., and Evgenii Z. Meilikhov. *Handbook of Physical Quantities.* Boca Raton, FL: CRC Press, 1997. ISBN: 0849328616.

Woan, Graham. *The Cambridge Handbook of Physics Formulas.* Cambridge: Cambridge University Press, 2000. ISBN: 0521573491.

BIOGRAPHY

Leiter, Darryl J., and Sharon L. Leiter. *A to Z of Physicists.* New York: Facts on File, 2003. ISBN: 0816047987.

WEB SITES

Eric Weisstein's World of Physics: http://scienceworld.wolfram.com/physics/

Physics Central: http://www.physicscentral.com/

The Physics Laboratory, National Institute of Standards and Technology (NIST): http://www.nist.gov/physlab/what-we-do.cfm

Physics World has news, views, and information for the global physics community from IOP Publishing: http://physicsworld.com/

MISCELLANEOUS

Bloomfield, Louis A. *How Things Work: The Physics of Everyday Life,* 3rd ed. Hoboken, NJ: Wiley, 2006. ISBN: 047146886X.

Gundersen, P. Erik. *The Handy Physics Answer Book.* Detroit: Visible Ink Press, 1999. ISBN: 1578590582.

Zoology

GUIDE TO THE LITERATURE

Schmidt, Diane. *Guide to Reference and Information Sources in the Zoological Sciences.* Westport, CT: Libraries Unlimited, 2003. ISBN: 1563089777.

DICTIONARIES

Allaby, Michael, editor. *A Dictionary of Zoology,* 3rd ed. New York: Oxford University Press, 2009. ISBN: 0199233403.

Barrows, Edward M. *Animal Behavior Desk Reference: A Dictionary of Animal Behavior, Ecology, and Evolution,* 2nd ed. Boca Raton, FL: CRC Press, 2001. ISBN: 0849320054.

Mcfarland, David. *Dictionary of Animal Behaviour.* New York: Oxford University Press, 2006. ISBN: 0198607210.

ELECTRONIC DATABASES

Biosis Previews (1926 to the present). Biosis Previews is a key database for literature searching in biological and life sciences. Covers medical topics with focus on basic research and its coverage of clinical medicine.

ISI Web of Science (1945 to the present, coverage varies). Institute for Scientific Information (ISI) http://scientific.thomson.com/isi/. A multidisciplinary database, with searchable author abstracts, covering the literature of the sciences, social sciences, and the arts and humanities. Users may choose to search across all the databases, or select just one. This unique database indexes and links cited references for each article. All formats provide complete bibliographic data and additional features, such as cited reference searching, links to related articles, plus author and publisher addresses.

Zoological Record Plus, ZR Plus (1878 to the present, produced by Thomson Scientific, and formerly by BIOSIS and the Zoological Society of London). Provides extensive coverage of the world's zoological and animal science literature, covering all research from biochemistry to veterinary medicine. The database provides an easily searched collection of references from over 5,000 international serial publications, plus books, meetings, reviews, and other nonserial literature from over 100 countries.

ENCYCLOPEDIAS

Bekoff, Marc, ed. *Encyclopedia of Animal Behavior.* Westport, CT: Greenwood Press, 2004. 3 vols. ISBN: 0313327459.

Bekoff, Marc, ed. *Encyclopedia of Human-Animal Relationships: A Global Exploration of Our Connections with Animals.* Westport, CT: Greenwood Press, 2007. 4 vols. ISBN: 9780313334870.

Cobb, Allan B., ed. *Animal Sciences.* New York: Macmillan Reference USA: Gale Group/ Thomson Learning, 2002. 4 vols. ISBN: 0028655567.

Cooke, Fred, et al. *The Encyclopedia of Animals: A Complete Visual Guide.* Berkeley: University of California Press, 2004. ISBN: 0520244060.

Eberhart, George M. *Mysterious Creatures: A Guide to Cryptozoology.* Santa Barbara, CA: ABC-CLIO, 2002. 2 vols. ISBN: 1576072835.

Grzimek, Bernhard, et al. *Grzimek's Animal Life Encyclopedia,* 2nd ed. Detroit: Gale, 2003. 17 vols. ISBN: 0787653624.

Hoagstrom, Carl W., ed. *Magill's Encyclopedia of Science: Animal Life.* Pasadena, CA: Salem Press, 2002. 4 vols. ISBN: 1587650193.

Macdonald, David W., ed. *The Encyclopedia of Mammals,* 2nd ed. New York: Facts on File, 2006. 3 vols. ISBN: 0816064946.

Perrins, Christopher, ed. *Firefly Encyclopedia of Birds.* Buffalo, NY: Firefly Books, 2003. ISBN: 1552977773.

Resh, Vincent H., and Ring T. Cardé, ed. *Encyclopedia of Insects.* Boston: Academic Press, 2003. ISBN: 0125869908.

HANDBOOKS AND TABLES

Alsop, Fred, J., III. *Birds of North America.* New York: DK, 2001. ISBN: 0789480018.

Baughman, Mel M., ed. *Reference Atlas to the Birds of North America.* Washington, DC: National Geographic, 2003. ISBN: 0792233735.

Coleman, Loren, and Patrick Huyghe. *The Field Guide to Lake Monsters, Sea Serpents and Other Mystery Denizens of the Deep.* New York: Jeremy P. Tarcher/Putnam, 2003. ISBN: 1585422525.

Marshall, Steven A. *Insects: Their Natural History and Diversity: With a Photographic Guide to Insects of Eastern North America.* Buffalo, NY: Firefly Books, 2006. ISBN: 1552979008.

Podulka, Sandy, et al., eds. *Handbook of Bird Biology,* 2nd ed. Ithaca, NY: Cornell Lab of Ornithology, in association with Princeton University Press, 2004. ISBN: 093802762X.

Wernert, Susan J., ed. *Reader's Digest North American Wildlife.* Pleasantville, NY: Reader's Digest, 1998. ISBN: 0762100206.

Wilson, Don E., and Sue Ruff, eds. *The Smithsonian Book of North American Mammals.* Washington, DC: Smithsonian Institution Press, 1999. ISBN: 1560988452.

WEB SITES

ACRL Internet Resources: Zoology: http://www.ala.org/ala/mgrps/divs/acrl/publications/crlnews/2008/apr/itsazoo.cfm

BiologyBrowser: http://www.biologybrowser.org/

Smithsonian National Museum of Natural History, Mammal Species of the World: http://vertebrates.si.edu/mammals/msw/

Smithsonian National Museum of Natural History, World of Copepods: http://invertebrates.si.edu/copepod/

MISCELLANEOUS

Tulin, Melissa S. *Aardvarks to Zebras: A Menagerie of Facts, Fiction, and Fantasy about the Wonderful World of Animals.* Secaucus, NJ: Carol Publishing Group, 1995. ISBN: 0806515481.

Index

About the Authors

JAMES E. BOBICK is the former head of the Science and Technology Department at the Carnegie Library of Pittsburgh, Pennsylvania, and a past visiting lecturer in the School of Information Science at the University of Pittsburgh. Bobick holds an MSLS in library science and an MS in biology. He is the author of the *Science and Technology Desk Reference: Over 1,700 Answers to Frequently-Asked or Difficult-to-Answer Questions* and *The Handy Science Answer Book*.

G. LYNN BERARD is a principal librarian in the science libraries at Carnegie Mellon University in Pittsburgh, Pennsylvania, and an adjunct faculty member for the Department of Library Science at Clarion University of Pennsylvania. She is a Fellow of the Special Libraries Association and is a contributor/author of the engineering and technology section of the annual reference publication *Magazines for Libraries*.